International Conference on Bioinformatics & Computational Biology (BIOCOMP 2017)

Held at CSCE'17

Las Vegas, Nevada, USA
17 – 20 July 2017

Editors:

Hamid R. Arabnia
Quoc-Nam Tran
Mary Yang

ISBN: 978-1-5108-5386-7

Printed from e-media with permission by:

Curran Associates, Inc.
57 Morehouse Lane
Red Hook, NY 12571

proceedings.com

Some format issues inherent in the e-media version may also appear in this print version.

Printed by Curran Associates, Inc. (2018)

For permission requests, please contact CSREA Press
at the address below.

CSREA Press
ATTN: Hamid R. Arabnia
1060 Long Creek Drive
Watkinsville, GA 30677
USA

Phone: (706) 542-3480

hra@cs.uga.edu

Additional copies of this publication are available from:

Curran Associates, Inc.
57 Morehouse Lane
Red Hook, NY 12571 USA
Phone: 845-758-0400
Fax: 845-758-2633
Email: curran@proceedings.com
Web: www.proceedings.com

PROCEEDINGS OF
THE 2017 INTERNATIONAL CONFERENCE ON
BIOINFORMATICS & COMPUTATIONAL BIOLOGY

BIOCOMP 2017

Editors

Hamid R. Arabnia
Quoc-Nam Tran
Mary Yang

CSCE'17
July 17-20, 2017
Las Vegas Nevada, USA
americancse.org

This volume contains papers presented at The 2017 International Conference on Bioinformatics & Computational Biology (BIOCOMP'17). Their inclusion in this publication does not necessarily constitute endorsements by editors or by the publisher.

AMERICAN
COUNCIL ON
SCIENCE AND
EDUCATION

Foreword

It gives us great pleasure to introduce this collection of papers to be presented at the 2017 International Conference on Bioinformatics and Computational Biology (BIOCOMP'17), July 17-20, 2017, at Monte Carlo Resort, Las Vegas, USA.

An important mission of the World Congress in Computer Science, Computer Engineering, and Applied Computing, CSCE (a federated congress to which this conference is affiliated with) includes *"Providing a unique platform for a diverse community of constituents composed of scholars, researchers, developers, educators, and practitioners. The Congress makes concerted effort to reach out to participants affiliated with diverse entities (such as: universities, institutions, corporations, government agencies, and research centers/labs) from all over the world. The congress also attempts to connect participants from institutions that have* ***teaching*** *as their main mission with those who are affiliated with institutions that have* ***research*** *as their main mission. The congress uses a quota system to achieve its institution and geography diversity objectives."* By any definition of diversity, this congress is among the most diverse scientific meeting in USA. We are proud to report that this federated congress has authors and participants from 64 different nations representing variety of personal and scientific experiences that arise from differences in culture and values. As can be seen (see below), the program committee of this conference as well as the program committee of all other tracks of the federated congress are as diverse as its authors and participants.

The program committee would like to thank all those who submitted papers for consideration. About 65% of the submissions were from outside the United States. Each submitted paper was peer-reviewed by two experts in the field for originality, significance, clarity, impact, and soundness. In cases of contradictory recommendations, a member of the conference program committee was charged to make the final decision; often, this involved seeking help from additional referees. In addition, papers whose authors included a member of the conference program committee were evaluated using the double-blinded review process. One exception to the above evaluation process was for papers that were submitted directly to chairs/organizers of pre-approved sessions/workshops; in these cases, the chairs/organizers were responsible for the evaluation of such submissions. The overall paper acceptance rate for regular papers was 25%; 18% of the remaining papers were accepted as poster papers (at the time of this writing, we had not yet received the acceptance rate for a couple of individual tracks.)

We are very grateful to the many colleagues who offered their services in organizing the conference. In particular, we would like to thank the members of Program Committee of BIOCOMP'17, members of the congress Steering Committee, and members of the committees of federated congress tracks that have topics within the scope of BIOCOMP. Many individuals listed below, will be requested after the conference to provide their expertise and services for selecting papers for publication (extended versions) in journal special issues as well as for publication in a set of research books (to be prepared for publishers including: Springer, Elsevier, BMC journals, and others).

- *Prof. Abbas M. Al-Bakry (Congress Steering Committee); University President, University of IT and Communications, Baghdad, Iraq*
- *Prof. Nizar Al-Holou (Congress Steering Committee); Professor and Chair, Electrical and Computer Engineering Department; Vice Chair, IEEE/SEM-Computer Chapter; University of Detroit Mercy, Detroit, Michigan, USA*
- *Prof. Hamid R. Arabnia (Congress Steering Committee); The University of Georgia, USA; Editor-in-Chief, Journal of Supercomputing (Springer); Fellow, Center of Excellence in Terrorism, Resilience, Intelligence & Organized Crime Research (CENTRIC).*
- *Prof. Hikmet Budak; Professor and Winifred-Asbjornson Plant Science Chair Department of Plant Sciences and Plant Pathology Genomics Lab, Montana State University, Bozeman, Montana, USA; Editor-in-Chief, Functional and Integrative Genomics; Associate Editor of BMC Genomics; Academic Editor of PLosONE*
- *Prof. Dr. Juan-Vicente Capella-Hernandez; Universitat Politecnica de Valencia (UPV), Department of Computer Engineering (DISCA), Valencia, Spain*
- *Prof. Kevin Daimi (Congress Steering Committee); Director, Computer Science and Software Engineering Programs, Department of Mathematics, Computer Science and Software Engineering, University of Detroit Mercy, Detroit, Michigan, USA*

We would like to extend our appreciation to the referees, the members of the program committees of individual sessions, tracks, and workshops; their names do not appear in this document; they are listed on the web sites of individual tracks.

As Sponsors-at-large, partners, and/or organizers each of the followings (separated by semicolons) provided help for at least one track of the Congress: Computer Science Research, Education, and Applications Press (CSREA); US Chapter of World Academy of Science; American Council on Science & Education & Federated Research Council (http://www.americancse.org/); HoIP, Health Without Boundaries, Healthcare over Internet Protocol, UK (http://www.hoip.eu); HoIP Telecom, UK (http://www.hoip-telecom.co.uk); and WABT, Human Health Medicine, UNESCO NGOs, Paris, France (http://www.thewabt.com/). In addition, a number of university faculty members and their staff (names appear on the cover of the set of proceedings), several publishers of computer science and computer engineering books and journals, chapters and/or task forces of computer science associations/organizations from 3 regions, and developers of high-performance machines and systems provided significant help in organizing the conference as well as providing some resources. We are grateful to them all.

We express our gratitude to keynote, invited, and individual conference/tracks and tutorial speakers - the list of speakers appears on the conference web site. We would also like to thank the followings: UCMSS (Universal Conference Management Systems & Support, California, USA) for managing all aspects of the conference; Dr. Tim Field of APC for coordinating and managing the printing of the proceedings; and the staff of Monte Carlo Resort (Convention department) at Las Vegas for the professional service they provided. Last but not least, we would like to thank the Co-Editors of BIOCOMP'17: Prof. Hamid R. Arabnia, Prof. Quoc-Nam Tran, and Prof. Mary Yang.

We present the proceedings of BIOCOMP'17.

Steering Committee, 2017
http://americancse.org/

Contents

SESSION: BIOINFORMATICS, NOVEL ALGORITHMS AND APPLICATIONS

SESSION: COMPUTATIONAL BIOLOGY, NOVEL ALGORITHMS, APPLICATIONS, AND TOOLS

SESSION: LATE PAPERS - COMPUTATIONAL BIOLOGY

SESSION

BIOINFORMATICS, NOVEL ALGORITHMS AND APPLICATIONS

Chair(s)

TBA

Genomic data mining reveals a rich repertoire of transporters in pathogenic fungi *Fusarium*

Abdulwahab Alghazali[1,*], Hong Cai[2,*], Yufeng Wang[2], Jianying Gu[1]
[1]Department of Biology, College of Staten Island, CUNY, Staten Island, NY 10314, USA
[2]Department of Biology and South Texas Center for Emerging Infectious Diseases, University of Texas at San Antonio, San Antonio, TX 78249, USA. *equally contributed

Abstract - *Fusarium is a genus of filamentous ascomycete fungi that includes many plant pathogens of agricultural importance. Fusarium can cause rots, wilts, blights in host plants, leading to huge economic loss. Fusarium can also produce a wide array of secondary metabolites, some of which are toxic and can contaminate crop products. Transporters are believed to play important roles in the life cycle of Fusarium. Our comparative genomic analyses of nine complete Fusarium species revealed a rich repertoire of transport proteins, which play important roles in nutrient uptake and second metabolite secretion.*

Keywords: *Fusarium*, Transporter, Secondary metabolites, Comparative genomics, TMSs

"Regular Research Paper"

1 Introduction

Fusarium is a genus of filamentous fungi that contains many plant pathogens, opportunistic human pathogens, and toxic producers [1]. Pathogenic *Fusarium* show significant differences in host adaptation and specificity. *F. graminearum* (*Fg*, teleomorph: *Gibberella zeae*) has a narrow host range, infecting predominantly the cereals. It causes *Fusarium* head bright on wheat, barley, and rice, stalk rot and ear rot in corn, and seedling blight on corn and wheat [2, 3]. *F. virguliforme* (*Fv*) causes sudden death syndrome of soybean [4]. Some *Fusarium* species have a broad host range, infecting both monocotyledons and dicotyledons. Members of *F. oxysporum* (*Fo*) species are causative agents of wilt diseases in over one hundred plant species of horticultural, agricultural or forest importance [5].

Fusarium is well known for its powerful secondary metabolism system. Secondary metabolites (SMs) are small organic molecules that are not essential for normal cell growth, but they may provide selective advantage. *Fusaria* can produce a diverse variety of secondary metabolites (mycotoxins) that may contaminate the harvested grain and can affect human and animal health if they enter the food chain [7]. Thus, the effective management practices to control *Fusarium* pathogens are urgently needed.

Transporters are vitally important to all living organisms by enabling metabolism, biological synthesis and reproduction, intercellular communication, and environmental sensing. Transporters function in acquisition of nutrients and vitamins from the environment, production of metabolites, maintenance of ion homeostasis, release and translocation of macromolecules including signaling molecules and membrane proteins, and efflux of medicine and the toxins [8]. The International Transporter Consortium (ITC) has pointed out that transporters are clinically important in drug absorption and disposition, metabolism and excretion, and a thorough understanding of the mechanism of transporters is critical for drug development [9-10].

In fungi, genes responsible for the synthesis of secondary metabolites (SMs), which cause crop grain contamination, are usually physically arrayed in a biosynthetic gene cluster. SM biosynthetic gene clusters also include genes that encode: (1) enzymes that tailor the precursor or intermediates in the pathway; (2) transporters that move SMs or intermediates across membranes; and (3) pathway-specific transcription factors [11-12]. The capacity to secret endogenous toxins favors the survival of fungi to competing organisms or host organisms. As the causative agent of "bakanae" disease of rice, *F. fujikuroi* is best known for its ability to produce gibberellins (GAs). The recent genome-wide analyses revealed that GA biosynthesis is limited to *F. fujikuroi*, suggesting that the diversity of SMs provides a selective advantage during infection of the preferred host plant rice and likely contributes to its adaptation to environmental changes [13]. In plant pathogens, transporters play an essential role in protection against plant host defense compounds during pathogenesis. The whole genome sequence of the cereal pathogen *F. graminearum* showed that other than transcription factors and hydrolytic enzymes, transmembrane transporters are major pathogenicity-related protein families [14-15]. Membrane-bound transporter can also affect the sensitivity of fungal pathogens to antifungal drugs such as azoles [16].

The availability of genomes from closely related *Fusarium* species enables comprehensive analysis of the transporter protein families in *Fusarium*. Comparative genomic analysis of three *Fusarium* species (*Fg, Fv*, and *Fo*) identified a total of 46 secondary metabolite biosynthesis (SMB) gene clusters [8]. In this study, we reported a catalog and comparative genomic analysis of transporters in nine *Fusarium* species with complete genomes and annotations.

Table 1: Genomic data for ***Fusarium*** species

Species	Accession ID (RefSeq / INSDC)	Genome size (Mb)	No. proteins	No. Transporters	% Transporters
F. avenaceum (Fav)	JQGD01000001 : JQGD01000105	42.71	13,092	755	5.8%
F. fujikuroi (Ff)	HF679023.1 : HF679034.1	43.65	14,813	824	5.6%
F. graminearum (Fg)	NC_026474.1 : NC_026477.1	36.45	13,313	705	5.3%
F. langsethiae (Fl)	JXCE00000000.1	37.54	11,940	668	5.6%
F. oxysporum (Fo)	NC_030986.1 : NC_031000.1	51.76	21,123	971	4.6%
F. poae (Fp)	LYXU01000001 : LYXU01000181	46.48	14,740	719	4.9%
F.pseudograminearum (Fps)	NC_031951.1 : NC_031954.1	36.97	12,447	684	5.5%
F. solani (Fs)	MSJJ00000000.2	51.32	15,708	955	6.1%
F. verticillioides (Fv)	NC_031675.1 : NC_031685.1	41.84	16,115	845	5.2%

We identified and classified these transporters, using the nomenclature in the transporter classification database TCDB. An improved understanding of *Fusarium* transporters will bring new insight into the mechanism underlying the pathogenesis and unique secretion systems of secondary metabolites in this group of fungus of enormous economic and agricultural significance.

2 Methods

2.1 Data

The completed whole genome data of nine *Fusarium* species (Table 1), including protein sequences and functional annotations were downloaded under bioprojects from NCBI Genbank database (https://www.ncbi.nlm.nih.gov/genome/). The curated transporters, including protein sequence, classification, structural, functional and evolutionary information about transport systems from a variety of living organisms were retrieved from the transporter classification database TCDB (http://tcdb.org/) [17-18] and TransportDB 2.0 (http://www.membranetransport.org) [19].

2.2 Identification and classification of transporters

The BLASTP search of all the proteins in nine *Fusarium* species versus all transporters in TCDB database were conducted to identify potential *Fusarium* transporters that are homologous to any known or predicted transporters in TCDB. The following criteria were used to define homologous genes: Expect value (E) $\leq$ 1e-5, sequence similarity $\geq$ 50%, and sequence coverage $\geq$ 30%. The classification of a *Fusarium* transporter was based on the known function of its homologous match that had the lowest expect value, the highest similarity score and the highest sequence coverage in the TCDB. The Pfam search based on the Hidden Markov Models (HMMs) was be used to identify conserved structural domains in transporters, using Pfam GA as the threshold [20]. Transmembrane protein topology and the number of putative TMSs were be predicted by TMHMM (http://www.cbs.dtu.dk/services/TMHMM/) [21]. The classification of fungal transporters in the TransportDB, the annotations and the conserved functional domain information was used to help filter false negative and false positive predictions.

Based on the degree of similarities with known or predicted transporters in TCDB, as well as the conserved domains and TMSs, we further classified the predicted transporters into classes, subclasses, families, and subfamilies according to the TC system. Transport Commission (TC) system developed by Saier group (http://www.tcdb.org/) is a comprehensive hierarchical classification system for membrane transport proteins [17-18].

3 Results and Discussion

We used the coding sequences from nine *Fusarium* genomes to query the transporter classification database TCDB (http://tcdb.org/) [17-18] using BLASTP and identified 668-1394 transporters in these genomes, which account for 4.9 to 6.1% of each respective proteome (Table 1). *F. oxysporum*, which has the largest genome, and the greatest number of protein-encoding genes, has the greatest number of transporters, whereas *F. langsethiae* contains only 668 transporters, the lowest number of transporters among the nine *Fusarium* species. Overall, all nine *Fusarium* species show the similar proportion of transporters in their proteomes.

Table 2. Distribution of topological types of transporters in nine *Fusarium* genomes

TMS	*Fav*	*Ff*	*Fg*	*Fl*	*Fo*	*Fp*	*Fps*	*Fs*	*Fv*
0	23	25	26	25	58	26	26	28	28
1	29	34	30	32	64	27	29	34	33
2	18	12	23	20	33	22	23	20	18
3	25	24	26	29	59	23	23	26	34
4	41	35	29	31	50	36	30	35	33
5	28	32	37	32	71	29	31	35	38
6	28	27	36	29	90	24	27	37	33
7	46	52	48	44	121	54	46	53	56
8	34	40	35	39	93	40	38	59	42
9	52	55	42	45	95	47	35	54	53
10	104	100	91	76	158	73	79	129	99
11	100	116	86	91	188	107	92	128	109
12	167	205	149	125	238	147	151	238	207
13	17	23	18	16	22	19	21	25	20
14	29	34	21	18	36	31	23	40	29
15	8	4	5	6	5	8	4	9	6
16	3	3	1	6	8	3	2	3	3
17	0	1	1	2	1	1	2	1	2
18	0	1	0	0	0	0	0	0	1
19	1	0	0	0	2	0	0	0	0
20	1	0	0	0	0	0	0	0	1
21	0	0	0	0	0	0	0	1	0
22	0	1	0	1	2	0	0	0	0
23	0	0	0	0	0	0	1	0	0
24	1	0	1	1	0	1	1	0	0
25	0	0	0	0	0	1	0	0	0

3.1 *Fusarium* transporters show diverse transmembrane topology

The capacity of a transporter is often associated with the complexity and topology of its transmembrane region(s) where the major events of substrate uptake or output across the cell membrane take place. Using the TMHMM (TransMembrane prediction using Hidden Markov Models) algorithm [21], we performed the transmembrane topology analysis for *Fusarium* transporters to identify the transmembrane segments (TMSs). The number of TMSs observed in a transporter in the nine *Fusarium* genomes varies from 0 to 25. The largest number of TMSs observed in a transporter in these genomes varies from 20 to 25 (Table 2). Except for intro-/extra-cellular transporters which have no TMS, transporters with 10, 11 and 12 TMSs are predominant. Most of transporters with 10 TMSs are members of the Major Facilitator Superfamily (MFS) (TC 2.A.1), the P-type ATPase (P-ATPase) Superfamily (TC 3.A.3), the Amino Acid-Polyamine-Organocation (APC) Family (TC 2.A.3), and the Telurite-resistance/Dicarboxylate Transporter (TDT) Family (TC 2.A.16). Transporters with 11 and 12 TMSs are mainly members of the Major Facilitator Superfamily (MFS) (TC 2.A.1), the Amino Acid-Polyamine-Organocation Superfamily (APC) (TC 2.A.3), and the ATP-binding Cassette (ABC) Superfamily (TC 3.A.1).

3.2 Transporters in nine *Fusarium* genomes can be divided into 7 classes and 158 families

The *Fusarium* transporters fall into seven classes and transporter families according to the TCDB system (Table 3). The distribution of transporters are y similar across species.

The Electrochemical potential driven transporters (Class 2) is the most abundant class of transporters in *Fusarium*, which includes 439-692 transporters (representing about 65.5% to 72.5% of the total transportomes). The transporters in this class include uniporters, symporters, and antiporters. The most abundant family, MFS, in Class 2 transporters has been implicated in efflux of drug and plant defense compounds in fungi. Drug:H^+ Antiporter (DHA) family of the MFS has been shown to play important role in antifungal drug resistance [22].

Class 3 transporters, the Primary Active Transporters, are also widely found in *Fusarium*. 99-120 transporters in nine *Fusarium* genomes belong to this class, which accounts for 12.0% to 15.6% of all transporters. This class of transporters play important roles in various aspects of life cycle, especially in the uptake and efflux of secondary metabolites, cation transportation, and multidrug resistance.

Table 3 The Distribution of putative transporters of nine *Fusarium* species in each TC class and subclass.

Class Subclass	*Fav*	*Ff*	*Fg*	*Fl*	*Fo*	*Fp*	*Fps*	*Fs*	*Fv*
1. Channels/pores	**52**	**52**	**53**	**46**	**65**	**49**	**50**	**46**	**56**
	6.9%	6.3%	7.5%	6.9%	6.7%	6.8%	7.3%	4.8%	6.6%
1.A: α Type Channels	44	44	45	39	53	42	42	39	47
1.B: β-Barrel Porins	2	1	1	1	1	1	1	1	1
1.C: Pore-Forming Toxins (Proteins & Peptides)	1	1	1	1	5	1	1	2	2
1.F: Vesicles Fusion pores	1	1	1	1	1	1	1	0	1
1.H: Paracellular channels	1	1	1	1	1	1	1	2	1
1.I: Membrane-bounded channels	3	4	4	3	4	3	4	2	4
2. Electrochemical potential driven transport	**519**	**584**	**463**	**439**	**680**	**473**	**448**	**692**	**601**
	68.7%	70.9%	65.7%	65.7%	70.0%	65.8%	65.5%	72.5%	71.7%
2.A: Porters (uniporters, symporters, antiporters)	519	584	463	439	680	473	448	692	601
3. Primary Active transport	**100**	**102**	**102**	**104**	**121**	**107**	**99**	**120**	**101**
	13.2%	12.4%	14.5%	15.6%	12.5%	14.9%	14.5%	12.6%	12.0%
3.A: P-P-bond hydrolysis driven transport	91	92	91	95	110	97	88	109	91
3.D: Oxidoreduction-driven transporters	6	7	8	6	6	7	8	8	7
3.E: Light absorption-driven transporters	3	3	3	3	5	3	3	3	3
4. Group Translocation	**7**	**7**	**8**	**6**	**9**	**7**	**6**	**9**	**10**
	0.9%	0.8%	1.1%	0.9%	0.9%	1.0%	0.9%	0.9%	1.2%
4.C: Acyl CoA ligase-coupled transporters	0	1	1	1	2	1	0	1	1
4.D: Polysaccharide Synthase/Exporters	2	2	3	2	3	2	3	3	5
4.E: Vacuolar Polyphosphate Polymerase-catalyzed Group Translocators	2	2	2	2	2	2	2	2	2
4.F: Choline/Ethanolamine Phosphotransferase 1	3	2	2	1	2	2	1	3	2
5. Transmembrane electron carrier	**2**	**2**	**4**	**3**	**3**	**3**	**3**	**2**	**3**
	0.3%	0.2%	0.6%	0.4%	0.3%	0.4%	0.4%	0.1%	0.4%
5.B Transmembrane one-electron transfer carriers	2	2	4	3	3	3	3	2	3
8. Accessory Factors in transport	**9**	**10**	**11**	**10**	**10**	**11**	**11**	**11**	**10**
	1.2%	1.2%	1.6%	1.5%	1.0%	1.5%	1.6%	1.2%	1.2%
8.A: Auxiliary transport proteins	9	10	10	10	10	10	10	11	10
8.B: Ribosomally synthesized protein / peptide toxins/agonists that target channels & carriers	0	0	1	0	0	1	1	0	0
9. Incomplete characterized transport system	**66**	**67**	**64**	**60**	**83**	**69**	**67**	**75**	**64**
	8.7%	8.1%	9.1%	9.0%	8.5%	9.6%	9.8%	7.9%	7.6%
9.A: Recognize transporters of unknown biochemical mechanism	29	32	26	24	40	27	25	39	29
9.B: Putative transport proteins	37	35	38	36	43	42	42	36	35

Class 1 transporters are not as abundant as Class 2 and 3 transporters, but are functional important for *Fusarium*. 46-65 channel/pores transporters are present in these nine genomes, accounting for 4.8% to 7.5% of all the transporters. The majority of these channel-type proteins are alpha-type channels (TC 1.A) which catalyze movement of solutes by passage through a transmembrane aqueous pore or channel. A small number of proteins belong to membrane-bounded channel, the Nuclear Pore Complex (NPC) (TC 1.I.1), which is also present in yeast and vertebrates. In yeast, approximately 30 nucleoporins assemble to form a large multiprotein complex and regulate nucleocytoplasmic transport of macromolecules [23].

Class 4, 5, and 8 are relatively less abundant. The number of transporters in these classes is highly conserved among all nine *Fusarium* genomes investigated. A significant number of transporters can be grouped into Class 9, an incompletely characterized class. While their exact physiological roles are yet to be elucidated, they might be involved in the transport of lipid, implicated by their sequence similarities with the member of the Lipid-translocating

Exporter (LTE) Family (9.A.26). Widespread use of azoles has led to the rapid development of drug resistance in fungi. Rta2p, an LTE member transporter in *Candida albicans* was shown to be involved in calcineurin-mediated azole resistance and dihydrosphingosine transport [24].

3.3 Examples of important transporter families

Many of the 58 transporter families are involved in the transfer of ions, saccharides, amino acids, polypeptides, proteins, drugs, toxins and other compounds. The most abundant and perhaps most important families are in the MFS superfamily (TC 2.A.1), the APC superfamily (TC 2.A.3), the Mitochondrial Carrier (MC) family (TC 2.A.29), the Drug/Metabolite Transporter (DMT) superfamily (TC 2.A.7), and the ABC superfamily (TC 3.A.1).

3.3.1 The MFS transporters

30.8% - 40.7% of all the transport proteins in the *nine Fusarium* genomes are members of MFS superfamily. The MFS transporters contain 12 or 14 transmembrane-spanning helices and drive translocation of the substrate by the proton concentration gradient generated across the plasma membrane. This superfamily is divided into 74 distinctive families based on phylogeny and function and transports a wide array of substrates, including sugars, drugs, metabolites, amino acids, nucleosides, vitamins, and a large variety of anions and cations [25]. Most MFS transporters in *Fusarium* are members of the following families:

(1) The sugar porter (SP) family (TC 2.A.1.1): this is the largest family. Members of this family usually have 12 established or putative TMSs. This family of transporters is very diverse in substrates and transport strategy. It has been shown that some *F. graminearum* sugar transporter genes are preferentially expressed in specific hosts, suggesting that *Fusarium* use host-specific gene expression of SP family members to modulate its primary response to a wide range of host plants and the uptake of plant-produced nutrients [26].

(2) The anion:cation symporter (ACS) family (TC 2.A.1.14), is a relatively large family with 37-95 members in *Fusarium* genomes. Inorganic anion symporter usually cotransport Na^+, while organic anion symporter cotransport H^+. The presence of 7-14 copies of Allantoate permease (TC 2.A.1.14.4) with dipeptide transporter activity may reflect that *Fusarium* might be dependent from the host plant for nitrogen supply.

3.3.2 The ABC transporters

About 5.4-7.6% (45-61) of the predicted *Fusarium* transporters are ATP-dinging cassette (ABC) transporters. ABC superfamily, along with MFS, are the two largest superfamilies of transmembrane transporters found in nature. Unlike MFS, ABC transporters are characterized by a conserved ATP-hydrolyzing domain to provide energy for translocation. All nine *Fusarium* species have a considerably higher number of ABC transporters than other fungi. The larger number of ABC transporters may account for the increase in host range of *F. oxysporum* as ABC transporters may provide tolerance to antifungal compounds generated by different hosts [27].

4 Conclusions

Comparative genomic analyses of nine *Fusarium* genomes revealed a rich repertoire of 668-971 transporters, belonging to seven transporter classes and 58 transporter families. The powerful transporter systems in *Fusarium* play critical roles in antifungal drug efflux, protein and second metabolite secretion, and stress response for host invasion. A better understanding of transport systems will allow enhanced optimization for disease control in economic important crop industry.

5 References

[1] George Agrios. "Plant Pathology". 5th edition Academic Press, 2005

[2] Goswami RS, Kistler HC. "Heading for disaster: *Fusarium graminearum* on cereal crops"; Mol Plant Pathol. 5(6): 515-525. Nov 2004.

[3] Dal Bello GM, Monaco CI, Simon MR. "Biological Control of seedling blight of wheat caused by *Fusarium graminearum* with beneficial rhizosphere microorganisms"; World J Microb Biot., 18(7): 627-636, Oct 2002.

[4] Srivastava SK, et al. "The genome sequence of the fungal pathogen *Fusarium virguliforme* that causes sudden death syndrome (SDS) in soybean"; Plos One, 9(1): e81832, Jan 2014.

[5] Michielse CB, Rep M. "Pathogen profile update: *Fusarium oxysporum.*"; Mol Plant Pathol., 10(3): 311-324, May 2009.

[6] Howard DH. "Pathogenic Fungi in Humans and Animals" 2nd edition (Marcel Dekker), 2003.

[7] Nesic K, Ivanovic S, Nesic V. "Fusarial toxins: secondary metabolites of *Fusarium* fungi" ; Rev Environ Contam Toxicol., 228: 101-120, 2014.

[8] Busch W, Saier Jr MH. "The IUBMB-endorsed transport classification system"; Mol Biotechnol., 27(3): 253-262, Jul 2004.

[9] Kell DB and Oliver SG. "How drugs get into cells: tested and testable predictions to help discriminate between

transporter-mediated uptake and lipoidal bilayer diffusion"; Front Pharmacol., 5: 231, 2014.

[10] International Transporter Consortium, et al. "Membrane transporters in drug development"; Nat Rev Drug Discov., 9(3): 215-236, Mar 2010.

[11] Keller N, Hohn T. "Metabolic pathway gene clusters in filamentous fungi"; Fungal Genet Biol., 21(1):17-29, Feb 1997.

[12] Walton JD. "Horizontal gene transfer and the evolution of secondary metabolite gene clusters in fungi: An hypothesis"; Fungal Genetics and Biology, 30(3):167-171, Aug 2000.

[13] Wiemann P, et al. "Deciphering the cryptic genome: genome-wide analyses of the rice pathogen *Fusarium fujikuroi* reveal complex regulation of secondary metabolism and novel metabolites"; PLoS Pathog., 9(6): e1003475, 2013.

[14] Cuomo CA, et al. "The *Fusarium graminearum* genome reveals a link between localized polymorphism and pathogen specialization"; Science, 317(5843): 1400-1402, Sep 2007.

[15] Del Sorbo G, Schoonbeek H, De Waard MA. "Fungal transporters involved in efflux of natural toxic compounds and fungicides"; Fungal Genet Biol., 30(1): 1-15, Jun 2000.

[16] de Waard MA, Andrade AC, Hayashi K, Schoonbeek HJ, Stergiopoulos I, Zwiers LH. "Impact of fungal drug transporters on fungicide sensitivity, multidrug resistance and virulence"; Pest Manag Sci., 62(3): 195-207, Mar 2006.

[17] Saier MH Jr, Yen MR, Noto K, Tamang DG, Elkan C. "The Transporter Classification Database: recent advances"; Nucleic Acids Res., 37(Database issue): D274-278, Jan 2009.

[18] Saier MH Jr, Reddy VS, Tsu BV, Ahmed MS, Li C, Moreno-Hagelsieb G. "The Transporter Classification Database (TCDB): recent advances"; Nucleic Acids Res., 44 (D1): D372–379, Jan 2016.

[19] Elbourne LDH, Tetu S, Hassan K and Paulsen IT. "TransportDB 2.0: a database for exploring membrane transporters in sequenced genomes from all domains of life"; Nucleic Acids Res., 45 (D1): D320-D324, 2017.

[20] Finn RD, et al. "Pfam: the protein families database"; Nucleic Acids Res., 42(Database issue): D222-230, Jan 2014.

[21] Krogh A, Larsson B, von Heijne G, and Sonnhammer ELL. "Predicting transmembrane protein topology with a hidden Markov model: Application to complete genomes"; J Mol Biol., 305: 567-580, Jan 2001.

[22] Costa C, Dias PJ, Sá-Correia I, Teixeira MC. "MFS multidrug transporters in pathogenic fungi: do they have real clinical impact?"; Frontiers in Physiology, 5:197, May 2014.

[23] Grossman E1, Medalia O, Zwerger M. "Functional architecture of the nuclear pore complex"; Annu Rev Biophys., 41:557-84, 2012.

[24] Zhang, S.Q., Q. Miao, L.P. Li, L.L. Zhang, L. Yan, Y. Jia, Y.B. Cao, and Y.Y. Jiang. "Mutation of G234 amino acid residue in *candida albicans* drug-resistance-related protein Rta2p is associated with fluconazole resistance and dihydrosphingosine transport"; Virulence, 6: 599-607, 2015.

[25] Pao SS, Paulsen IT, Saier MH Jr. "Major facilitator superfamily"; Microbiol Mol Biol Rev., 62: 1–34, Mar 1998.

[26] Harris LJ, Balcerzak M, Johnston A, Schneiderman D, Ouellet T. "Host-preferential *Fusarium graminearum* gene expression during infection of wheat, barley, and maize"; Fungal Biol., 120(1):111-23, Jan 2016.

[27] Ma LJ, et al. "Comparative genomics reveals mobile pathogenicity chromosomes in *Fusarium*"; Nature, 464: 367-373, Mar 2010.

Advanced Agglomerative Clustering Technique for Phylogenetic Classification Using Manhattan Distance

Prof. Dr. Md Abdul Mottalib[1], **Raihan Islam Arnob**[1], **Md. Redwan Karim Sony**[1] **and Mrs. Lipi Akter**[1]
[1]Department of Computer Science and Engineering, Islamic University of Technology, Dhaka, Bangladesh

Abstract— *Classifying organisms on the basis of descenders from a common ancestor is called phylogenetic classification. In the field of bioinformatics, it exhibits an important role since it provides us with the hierarchical genetic information of a naturally cultured source. Metagenomic data is huge in size and of high dimension. To manipulate data of such high dimension, we require algorithms which can perform fast and provide us with meaningful information. Creating hierarchical tree for large number of genes does not provide that much of useful information since the level of the tree becomes enormous. The proposed technique* ***A****dvanced* ***A****gglomerative* ***C****lustering* ***T****echnique (AACT) consists of mainly two phases. At first it clusters the similar data using K-Means algorithm. Then centroid from each of those clusters is taken as the input of the next agglomerative hierarchal clustering phase. Our proposed method AACT using Manhattan Distance mainly aims to identify m number of distinct clusters over vast dataset with lower complexity and thus reducing the time complexity than existing methods.*

Keywords: S-Link, Manhattan Distance, Advanced Agglomerative Clustering Technique(AACT), Phylogenetic classification.

1. Introduction

Bio-informatics is an interdisciplinary field that develops methods and software tools for understanding biological data. As an interdisciplinary field of science, bioinformatics combines computer science, statistics, mathematics, and engineering in order to efficiently store, analyze and interpret biological data. Metagenomics[1] is a new field of study that provides us with a deeper insight into the microbial world compared to the traditional single genome sequencing technologies. Traditional methods for studying individual genomes are well developed. However, the are not appropriate for studying microbial samples from the environment because traditional methods depends on the clonal clusters cultivated in laboratory when more than 99% of bacteria are unknown and can not be cultivated and isolated[2]. Metagenomics use technologies that sequence uncultured bacteria genomes in an environment sample directly[3] and thus makes it possible to study organisms which can not be isolated or are difficult to grow in the lab.

There are several problems of metagenomic techniques. These are high computational complexity of large unstructured dataset obtained from the nature. Therefore in order to find meaningful information from this vast and unstructured data efficiently in reasonable time is necessary. From this perspective a clustering phase before classification of metagenomic sequences is necessary for grouping the similar data samples together in order to reduce the data redundancy and thus reduce the computational complexity. Since the amount of data is huge, then slight improvement in the computational complexity is amplified in a huge benefit in the final runtime of the data processing time.

The main goal of metagenomic study of any environmental samples is to identify the microorganisms and find out the interaction among them. In order to find out this relationship among different species in any environment we need to find out their evolutionary relationship. Therefore we have to find out the phylogenetic dependency tree. A phylogenetic tree or evolutionary tree is a branching diagram or "tree" showing the inferred evolutionary relationships among various biological species or other entities and their phylogeny based upon similarities and differences in their physical or genetic characteristics[4].

Recent development in phylogenetic classification has resulted in methods like improved agglomerative clustering technique[5] which takes into account the Euclidian distance and Complete link (C-LINK) method to find the distance. But calculating the Euclidian distance is computation intensive since it has to square the difference of every dimentsion and then square root after the summation.

The main purpose of using data clustering technique is to improve the performance of data access by summarizing the data objects into groups[6]. As the dataset size in metagenomics is huge with comparatively large dimensions, applying clustering steps as one of the preporcessing step yields much computational complexity improvement. A clustering method which requires less computational cost can be beneficial in general data mining and knowledge discovery, as well as in specific domains[7]. In clustering techniques when the distance calculation is done, the computational complexity is $O(N^2)$, where N is the number of data points[8], [9], [10]. The linkage method used in clustering method is one of the most important aspect of the performance of clustering. In this paper we make the use of *S-LINK (Single Link)* as the linking method which operates based on the minimum possible distance between the two clusters.

Microorganism can be found in almost every environment of the earth's biosphere and are responsible for numerous biological activities including carbon and nitrogen cycling [11], organic containment remediation[3], [12] and human health and disease. Many human disorders, such as type 2 diabetes (T2D), obesity, dental cavities, cancer and some immune-related diseases are known to be related with a single or group of microorganisms[13], [14], [15], [16], [17], [18]. In addition, different strains within the same species may have completely different impacts on human health, such as Escherichia Coli, which is highly virulent E. coli strain, whereas most other strains in this same species are non-pathogenic. Thus characterization and identification of microbial strains/species in the environment and individual human hosts is of crucial importance to reveal human-microbial interactions, especially for patients with microbial-mediated disorders.

This paper is organized as follows. In section 2, we discuss the proposed methodology with associated stages. Section 3 mainly discusses the experimental setup, experimental result and performance analysis. In Section 4, the conclusion of our study is described.

2. Proposed Method

The **A**dvanced **A**gglomerative **C**lustering **T**echnique **(AACT)** works in two stages. At first the traditional K-means algorithm is used to to group the similar data samples together in order to reduce the extra computations required to process similar data. Then representatives from each of the cluster or groups is taken out to use as the input dataset for the next agglomerative hierarchical clustering. The flowchart of the process is depicted in Figure (1).

In the flowchart in figure(1), at first the dataset is taken and normally dataset is in the form of natural logarithmic value of intensity. So they are converted to their original values with exponent. Then the number of clusters in the K-Means stage l is selected and thus with specifying the number of maximum iterations, K-Means clustering phase is completed. Next to this, representatives from each of the clusters is taken as an input for the agglomerative hierarchical clustering.

2.1 Proposed Algorithm

The **A**dvanced **A**gglomerative **C**lustering **T**echnique (AACT) using Manhattan Distance works in the following way.

2.1.1 K-means Stage

In this stage the traditional k-mean technique is applied and identified l distinct clusters over the input dataset X. Generally, the traditional k-means technique consists of three steps. In the first step, to fix the l centroids values $\bar{K} = \{\bar{K}_1,...,\bar{K}_l\}$ over the input dataset X as defined $X = \{X_1,..., X_n\}$, where X represents input dataset, n denotes number of

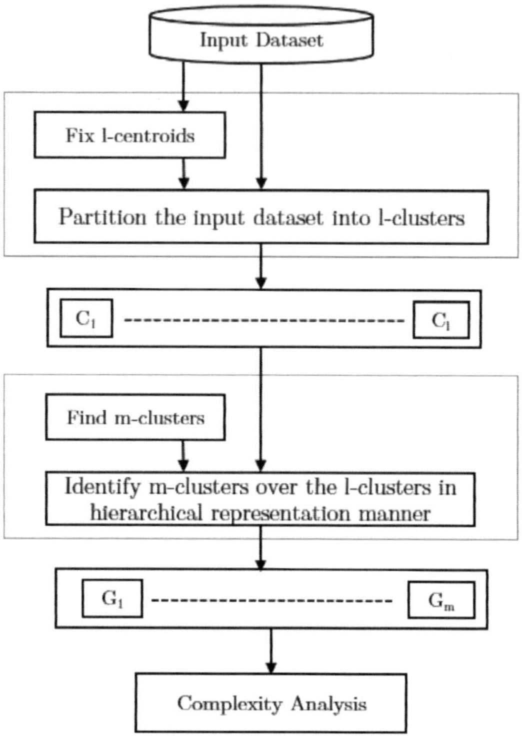

Fig. 1: Flowchart

objects that belong to input dataset X and $\bar{K}$ represents the number of centroid values identified in X. In the second step, it maps the l clusters in $\bar{K}$ over the input dataset X through the process of measuring Manhattan distance between dataset X and l centroid values as defined in the equation(1) in the follwoing.

$$C_j = Min\{D(X_i, \bar{K}_j) \mid \forall X_i \in X, \forall \bar{K}_j \in \bar{K}_l\} \quad (1)$$

where $D(X_i, \bar{K}_j)$ represents the Manhattan distance between i^{th} object in X and j^{th} centroid in $\bar{K}$ and is defined as equation (2)

$$D(X_i, \bar{K}_j) = \{abs((X_i - \bar{K}_j)) \mid \forall X_i \in X, \forall \bar{K}_j \in \bar{K}\} \quad (2)$$

where X_i denotes the dataset X and $\bar{K}_j$ is centroid value of jth cluster. Here the function *abs()* provides the absolutve value for the distance calculation in Manhattan distance. The Manhattan distance is based on absolute value distance, as opposed to squared error (Euclidean distance) distance. In practice, both of them provides nearly similar results most of the time. Absolute value distance should give more robust results, whereas Euclidean would be influenced by unusual values.

This is a multivariate technique, and "distance" between two points involves aggregating the distances between each variable. So if two points are close on most variables, but more discrepant on one of them, Euclidean distance will exaggerate that discrepancy, whereas Manhattan distance will shrug it off, being more influenced by the closeness of the other variables.

In the next step, it partitions the input dataset X into l distinct clusters $C = \{ C_1,...,C_l \}$ in $\bar{K}_j$ as defined in equation (3)

$$\bar{K}_{\rm j} = \{\frac{1}{N_{\rm j}}\sum_{l=1}^{n_{\rm j}} C_{\rm jl} \mid \forall C_{\rm jl} \in C_{\rm j},\ \forall C_{\rm j} \in C\} \tag{3}$$

Where C_{ij} represents the ith object in the jth cluster that belongs to the C. Repeat the steps from step 2 to step 3 until the result of the current iteration equal to previous iteration. This modified K-means algorithm is described in the below subsection.

2.1.2 Algorithm for K-means Clustering

This algorithm shows the basic operation procedure of K-means clustering which is used here in this paper as a first step of clustering before starting the hierarchical clustering.

Input: $X = \{X_1,..., X_n\}$
Output: l-clusters = $\{C_1,C_2,..., C_l\}$
Begin

1) Fix the l centroids values $\bar{K} = \{\bar{K}_1,..., \bar{K}_n\}$ over the input dataset X.
2) Map l clusters in $\bar{K}$ over the input dataset X by using the equation (1) and (2).
3) Partition the input dataset X into l distinct clusters C = $\{C_1,..., C_l\}$ using the equation(3).

End

2.1.3 S-LINK Stage

In this stage, the S-LINK (Single Linkage) technique is applied to calculate the distance taken in consideration between the two clusters as the distance between two clusters. This is a very important step in identifying *'m'* clusters over the result of k-means technique *C*. *S-LINK* technique consists of four steps. In the first step, it computes centroid over each individual clusters from the result of K-means algorithm applied in the previous step, C for $i=1,2,....,l$ using the equation(4).

$$\bar{C} = \sum_{i=1}^{l}\sum_{j=1}^{n_{\rm i}} C_{\rm ij} \tag{4}$$

Where C_{ij} denotes the jth object in the ith cluster, l denotes the number of clusters and n_i denotes number of objects in the ith cluster. In the second step, it constructs the distance matrix D_{ij} over the result of $\bar{C}$ based on Manhattan distance and is defined in equation (5).

$$D_{\rm ij} = \{_{\rm i=1,2,...k;\ j=\ i+1,...k}\ d(\bar{C}_{\rm i},\bar{C}_{\rm i}) \mid \forall \bar{C}_{\rm i} \in \bar{C},\ \forall \bar{C}_{\rm j} \in \bar{C},\} \tag{5}$$

Where $d(\bar{C}_{\rm i},\bar{C}_{\rm j})$ represents the distance between i^{th} and j^{th} cluster belonging in $\bar{C}$ and is defined in equation(6).

$$d(\bar{C}_{\rm i},\bar{C}_{\rm j}) = |\bar{C}_{\rm i} - \bar{C}_{\rm j}| \tag{6}$$

If i^{th} and j^{th} cluster are containing more than one objects, then compute the distance of set of objects then compute the distance of set of object pairs between i^{th} and j^{th} clusters and then consider the minimum distance of object pair as a distance of ith and j^{th} cluster as defined in equaion (7). This is the controlling stage of *S-LINK* Single Link stage as here minimum distance is selected as the distance between the two clusters.

$$d(\bar{C}_{\rm i},\bar{C}_{\rm j}) = min\{d(\bar{C}_{\rm i},\bar{C}_{\rm j})\} \tag{7}$$

where $\bar{C}_{\rm i}$,$\bar{C}_{\rm j}$ denotes object pairs of $i^{\rm th}$ and $j^{\rm th}$ clusters and $\bar{C}$.

In the fourth step, it finds the closest cluster pair with minimum distance Δd over the distance matrix $D_{\rm ij}$ as defined in equation(8).

$$\Delta d = min\{D_{\rm ij} \mid \forall D_{\rm ij} \in D\} \tag{8}$$

Now the Agglomerative hierarchical clustering begins. We merge the closest cluster pair $(\bar{C}_{\rm i},\bar{C}_{\rm j})$ into a single cluster $\bar{C}_{ij}$. Then delete the jth and compute the centroid of new cluster $\bar{C}_i$. Repeat the step two, until the number of iterations is satisfying *(l-m)* where *m* is the number of clusters. This modified S-LINK algorithm is described in the following section.

2.1.4 Algorithm for Agglomerative Clustering

Input: $C = \{C_1,...,C_l\}$
Output: $G = \{G_1,...,G_l\}$
Begin

1) Compute centroid over the each individual clustersinthe result of K-means, C for $i=1,2,...,l$ using the equation(4).
2) Construct distance matrix D_{ij} over the result of $\bar{C}$ based on Manhattan distance in equation (5) and (6).
3) if i^{th} and j^{th} clusters are containing more than one objects then compute the distance of set of object pairs between i^{th} and j^{th} clusters and consider the minimum distance of object pairs as a distance of ith and j^{th} cluster using equation(7).
4) Find the closest cluster pair with minimum distance Δd over the distance matrix D_{ij} using the equation(8).

5) Merge the closest pair $(\bar{C}_i,\bar{C}_j)$ into single cluster $\bar{C}_{ij}$. Delete the jth cluster and compute centroid of new cluster $\bar{C}_i$. Repeat the steps, until the number of iterations is satisfying *(l-m)*.

End

2.1.5 Simulation Technique

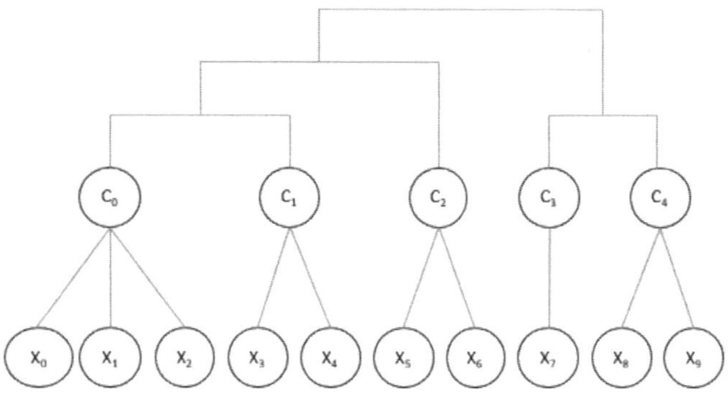

Fig. 2: Simulation Technique

Here in figure(2) let us consider $X_0,..., X_9$ are the preliminary dataset which is found out by exponential of the intensity values in the dataset. In the first step, the number of clusters is specified and $C_0,..., C_4$ are the random clusters of K-means method. Then rest of the data is assigned to the clusters based on the closest distance which is calculated basing Manhattan distance method.

Then from each of the clusters of K-means, one representative data is selected from each of the cluster. Now, in agglomerative stage, each pair of the closest cluster is merged together in each of the iteration and ultimately they merge to one cluster. If we want to get finally *m* number of clusters, then we have to stop *m* iterations before the iteration loop ends. Thus we get the phylogenetic tree.

3. Experimental Result and Performance Analysis

3.1 Complexity Analysis

The proposed Advanced Agglomerative Clustering Technique(AACT) is better suitable to identify *m* distinct clusters over the large dataset with lesser computational complexity and finite number of iterations. In stage one *(k-means)* dataset of size n is reduced to *l* distinct number of clusters with *l* number of iterations and computational complexity of $O(nl)$ where, *n* is the size of dataset and *l* is distinct number of clusters. In second stage *(S-LINK) l* distinct number of clusters obtained from stage one *k-means* is reduced to m number of groups with *(l-m)* number of iterations and computational complexity of $O(nl+l^2)$, where *nl* is the computational complexity of stage one and l^2 is the computational complexity obtained due to the construction of distance matrix D_{ij} in the S-LINK stage.

In this method, in case of using the distance function for distance matrix generation, using Manhattan distance rather than the Euclidian distance gives a consistent improvement in performance. Manhattan distance gives a good improvement in the performance than Euclidian distance as the Euclidian distance calculation requires squaring the values at each of the dimension and then adding them and squaring root. This squaring and subsequent square root gives a considerable calculation overhead for large dataset specially with higher data dimensionality like metagenomic data. Minkowski distance can also be used but they will give higher computational computational complexity due to calculation of higher power distance and subsequent higher square root.

3.2 Experimental Result

3.2.1 Dataset Description

- Title: Influenza virus H5N1 infection of U251 astrocyte cell line: time course
- Organism: *Homo sapiens*
- Platform: GPL6480: Agilent-014850 Whole Human Genome Microarray 4x44K G4112F (Probe Name version)
- Citation: Lin X, Wang R, Zhang J, Sun X et al[19]
- Sample Count: 6
- Value type: Intensity transformed count
- Published Date: 04-01-2016
- Summary: Analysis of U251 astrocyte cells infected with the influenza H5N1 virus for up to 24 hours. Results provide insight into the immune response of astrocytes to H5N1 infection.
- Weblink:NCBI database [1]

3.2.2 Test Bench

- Processor: Intel Core i7-5820K @ 3.3 Ghz
- Chipset : Intel X99 Express Chipset
- Ram: 8 GB @ 2400 Mhz
- Platform: MATLAB 2016 64 bit

3.2.3 Time Comparison

In this section we compare among the trivial agglomerative method, most recently developed Improved Agglomerative Clustering Technique and our proposed method. As we can see from the table that our proposed method works faster than the *IACT* as Manhattan Distance was used

[1] https://www.ncbi.nlm.nih.gov/

to calculate distance among clusters instead of Eucledian Distance, which reduced the time required to compute the phylogenetic classification. So, our proposed method works faster than the *IACT*.

Sample Count	IACT (sec)	AACT (Proposed) (sec)	Trivial Agglomerative (sec)
4000	20.4776	19.063	40.7660
8000	40.5365	38.4779	80.8364
12000	62.3473	56.6029	120.7839
16000	80.9419	76.1390	163.9884
20000	103.3372	97.3996	203.4975
24000	124.1642	115.2550	243.6152

Here we see from *Figure 5* that our proposed method is far better than trivial approach since we have shrinked the large dataset using K-means clustering which performs fast but is not a hierarchical approach. So, next we use S-Linkage agglomerative method to generate the hierarchical classification for this case our phylogenetic classification.

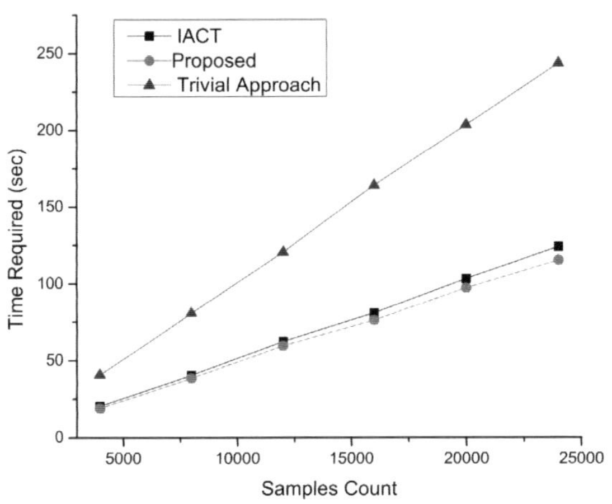

Fig. 3: Time Comparision

4. Conclusion

The size of dataset in Metagenomics is ridiculously large. In order to manipulate this vast and increasing dataset we need very efficient algorithms. Next, each of the cluster of **A**dvnaced **A**gglomerative **C**lustering **T**echnique should be annotated from the databank of *NCBI (National Center for Biotechnology Information)*.

So far, all the clustering algorithms runs in sequential execution but for further improvement these algorithms can be optimized for parallel execution. Even in this age of distributed computing, this system can easily be optimized for distributed systems.

References

[1] David Koslicki, Simon Foucart and Gail Rosen, "Quikr: a method for rapid reconstruction of bacterial communities via compressive sensing", *Advance Access Publication*, vol. 29, 2013, pp. 2096-2102.

[2] Genivaldo Gueiros Z. Silva, Daniel A. Cuevas, Bas E. Dutilh and Robert A. Edwards, "FOCUS: an alignment-free model to identify organisms in metagenomes using non-negative least squares", *PeerJ*, 2014.

[3] Turnbaugh, P.J., Hamady, M., Yatsunenko T. Cantarel, B.L, Duncan A, Ley R.E., Sogin, M.L., Jones, Roe, B.A. Affouritit, J.P. et al., "A core gut microbiome in obese and lean twins", *Nature*, vol. 457, 2009, pp. 480-484.

[4] https://en.wikipedia.org/wiki/Phylogenetic_tree

[5] Shreedhar Kumar S, Jithender M, Chaithra B, Md. Sharif Nawaz, Anushree D., "Improved Agglomerative Clustering Technique for Large Datasets", *25TH IRF INTERNATIONAL CONFERENCE*, 2016, pp. 4-8.

[6] Bhargavi M.S and Sahana D. Gowda, "An Hybrid Validity Index for Dynamic Cut-off in Hierarchical Agglomerative Clustering", *ICACCI Conference*, 2014.

[7] Athman Bouguettaya, Qi Yu, XuminLiu ,Xiangmin Zhou , Andy Song, "Efficient agglomerative hierarchical clustering", *Expert Systems with Applications*, vol. 42, 2015, pp. 2785-2797.

[8] Chih-Tang Chang ,JimZ.C.Lai , M.D.Jeng. "Fast agglomerative clustering using information of k-nearest neighbors", *Pattern Recognition*, vol. 43, 2010, pp. 3958-3968.

[9] Jim Z.C. Lai, Tsung-Jen Huang, "An agglomerative clustering algorithm using a dynamic k-nearest-neighbor list", *Information Sciences*, 2011.

[10] J. Shanbehzadeh and P. O. Ogunbona, "On the computational complexity of the LBG and PNN algorithmsâĂİ, *IEEE Trans Image Process*, vol. 6, 1997.

[11] https://en.wikipedia.org/wiki/Soil_biology

[12] N. Diaz L. Krause, A. Goesmann and et.al., "TACOA âĂŞ Taxonomic classification of environmental genomic fragments using a kernelized nearest neighbor approach", *BMC Bioinformatics*, 2009.

[13] D.L. Wheeler, T. Barreett, D. A. Benson, and et al. "Database resources of the National Center for Biotechnology Information", *Nucleic Acid Research*, vol. 35, 2007, pp. 5-12.

[14] D. A. Benson, I. Karsch-Mizrachi, D.J. Lipman, and et al, "IMGT, the international ImMunoGeneTics information system", *Nucleic Acids Research*, vol. 37, 2009.

[15] Ley, R.E. (2010) "Obesity and the human microbiome", *Curr Opin Gastroenterol*, 2010, pp. 5-11.

[16] Larsen, N., Vogensen, F.K., van den Berg, F.W.J, Nielson, D.S., Andersen, et al., "Gut Microbiota in Human Adults with Type 2 Diabetes Differs from Non-Diabetic Adults", *PLOS ONE*, 2010.

[17] S.D. Bently and J. Parkhill, "Comparative genomic structure of prokaryotes", *Annual Review of Genetics*, vol 38, 2004, pp. 771-791.

[18] Y. W. Wu and Y. Ye, "A novel abundance-based algorithm for binning metagenomic sequences using l-tuples", *Annual International Conference on Research in Computational Molecular Biology RECOMB'10*, 2010, pp. 535-549.

[19] Lin X, Wang R, Zhang J, Sun X et al. "Insights into Human Astrocyte Response to H_5N_1 Infection by Microarray Analysis", *Viruses*, 2015.

Sequence-based Deep Learning Reveals the Bacterial Community Diversity and Horizontal Gene Transfer

Hao Zheng[1], Jie Yin[2], Jieruo Gu[3*] and David Xingfei Deng[4*]

[1]School of Electrical and Computer Engineering, Georgia Institute of Technology, Atlanta, CA, USA
[2]Department of Horticulture and Landscape Architecture, Purdue University, West Lafayette, IN, USA
[3]Internal Medicine, The 3rd Hospital of Sun Yat-Sen University, Guangzhou, China 510630
[4]Ardent BioMed Guangzhou Inc., Guangzhou, China 510530
[*]Corresponding authors (daviddeng@ardentbiomed.com)

Abstract - Metagenomics is the application of advanced genomic techniques to the study of microbial communities in their native environments, and does not require lab cultivation nor isolation of individual genomes. Here, we propose a novel approach for metagenomics operational taxonomic unit (OTU) assignment using deep learning. The experimental results demonstrate that a PCA-based convolutional neural network is very powerful and fast for metagenomics OTU assignment and also prioritizing horizontal gene transfer events.

Keywords: OTU, metagenomics, sequencing, deep learning

1 Introduction

Metagenomics studies the whole microbial communities in their native environments by using advanced computational and genomic techniques, and does not need lab cultivation nor isolation of individual genomes. It is a very powerful method to reveal sample relative abundance and the genetic material of a microbe or entire communities of organisms directly from its environment without losing its nativity [1].

NGS-based metagenomics is very hot and has been popularly used in various studies. For example, human gut microbiomes were studied by using fecal samples from individuals of various ages through computational biology approaches, leading to the discovery of age-related DNA commonality and uncertainty [13][17]. Researchers also collected the Seawater samples from the Sargasso Sea and the diversity of microbial communities and gene signatures were analyzed through WGS sequencing [10]. This initial project was further expanded into the Sorcerer II Global Ocean Sampling expedition, in which a total of 44 samples were gathered from the Northwest Atlantic through Eastern Tropical Pacific and sequenced using NGS-based biological approaches to study microbial genomes and the richness in the surface seawater [11][12]; Approaches alike were also explored to study the bacteria lives and their functional potentials in vagina, human saliva, human skin, and soils [6]-[9][14][15].

To determine both the community composition and the potential physiology of the abundant community members in the metagenomics sample, OTU assignment is a critical step [23]-[25]. Metagenomic OTU assignment is usually done based on the contigs or unassembled reads from the sequencing data to cluster these contigs or reads into closely related populations [27]. Depending on the structure of the microbial community, the quality and depth of reads, OTU assignment can be performed at different levels, e.g. to classify reads of the same superkingdom into high levels like phyla or even down to the finer species levels.

For such OTU assignment, a number of *in silico* methodologies have been devised recently. These methods can be generally split into two main categories: (a) supervised methods and (b) unsupervised methods [16]. Most supervised methodologies are relying on some sort of taxonomical information. To that end, reference databases are usually needed for the assigning taxonomical levels to contigs or reads. The intrinsic algorithm utilizes either sequence similarity, or genomic signatures such as tandem nucleotide sequencing composition patterns. Such methods inlcudes MEGAN, Phylopythia, NBC, PhymmBL, and SPHINX [18]-[22].

In comparison, clustering approaches are usually independent of taxonomy. Such methods generally requires no additional reference databases nor taxonomic information. The basic idea behind taxonomy independent methods is that reads from different species may have some intrinsic patterns. For instance, different α-proteobacteria species may have guanine-cytosine contents spanning <30% to >60% [28]. Popular methods under this category include TETRA, variants of SOMs, CompostBin, AbundanceBin and MetaCluster [29]-[33].

2 Materials and Methods

2.1 Data Sets

Because of the nature of metagenomics, we don't have the truth for OTU. To benchmark the performance, we used in-house simulated data sets that were generated from available bacterial genomes. The dataset is further partitioned to serve as training and testing based on cross validation schema to be described below.

We also tested the performance of the proposed method on a very comprehensive synthetic metagenomic data set named simHC [26]. SimHC simulated a high-complexity microbial community by the Integrated Microbial Genomes and Microbiomes system of JGI. The lengths of the genomic fragments in the simHC data set span from 130 up to 3,754 bps. And, about 2.6% of nucleotides in simHC were not specified, mimicking the noisiness encountered during next generation sequencing.

2.2 Preprocessing

The raw metagenomics data is usually noisy and may contain DNA sequences of eukaryotic origin. Preprocessing of the raw data is a critical upstream step to ensure meaningful prefiltering includes removal of redundant, low quality sequences and sequences of eukaryotic origin [3][4]. Oligo frequencies patterns up to hexamer are extracted. As the pattern often contains redundant information, mapping it to a feature vector can get rid of this redundancy and yet preserve most of the intrinsic information content of the pattern. These extracted features have great role in distinguishing input patterns.

2.3 Deep Learning

We used principal component analysis (PCA) for extracting the features to serve as the inputs to deep learning. Based on the eigenvectors produced by the PCA methods, the projection vectors from the training set were obtained and then serves as the inputs to the neural network. Usually, we need a lot of training epochs to learn meaningful weights, or we require related data sets to be used for seeding a fine-tuning of transfer learning network. Here, we turn a PCA into an auto-encoder, by generating an encoder level of the PCA parameters and furthermore adding a decoder level [34].

To extract meaningful and representative features from a high dimensional space is usually challenging. Such a problem is well known as curse of dimensionality. To address this problem, we utilized PCA transformation to serve as the inputs to the neural network [2]. More specifically, we then set the weight matrix of the neutral network as Θ and as a result the initialization cost of the network depends only on the number of samples serving to obtain the principal components.

Relu activation function was used over traditional sigmoid and tanh fucntion. The Relu function is defined as follows

$$f(x) = \max(0, x) \quad (1)$$

Relu activation function has its unique advantages over traditional sigmoid and tanh function. The training of the network using Relu is much faster, with reduced likelihood of the gradient to vanish. Also, sparse representations resulting from the Relu activation fucntion is normally more beneficial than dense representations.

2.4 Training Schema

57% of the samples are used for training. These are presented to the network during training, and the network is adjusted according to its error. 10% of the samples are used for validation. These are used to measure network generalization, and to halt training when generalization stops improving. The rest 33% of the samples are reserved for testing. These have no effect on training and so provide an independent measure of network performance during and after training. Training will stop by itself at the time generalization stops getting improving. This is indicated by an increase in the cross-entropy error of the validation sample set.

3 Results

The results demonstrate that this sequence-based deep learning method can reveal the bacterial community diversity with high accuracy and indicate potential underlying horizontal gene transfer events.

3.1 Performance

Figures 1-2 illustrate that the performance of the proposed method outperforms two popular existing methods (TETRA and Phylopythia) using both in house simulated datasets and widely used simHC data set in terms of specificity and sensitivity.

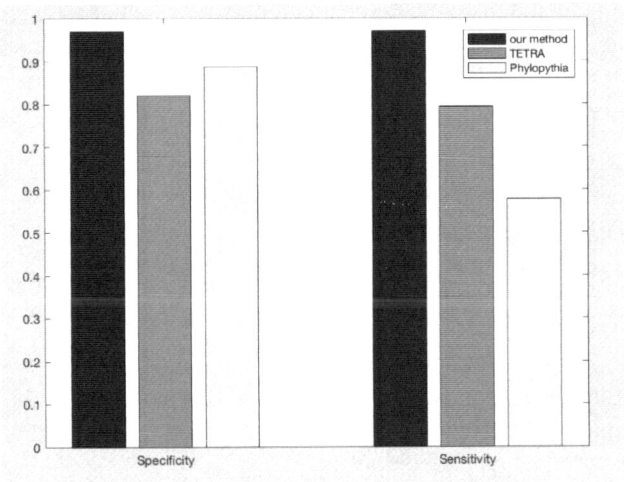

Figure 1. Comparison of Specificity and Sensitivity of the proposed method versus popular existing methods at the superkingdom level using simulated metagenome data.

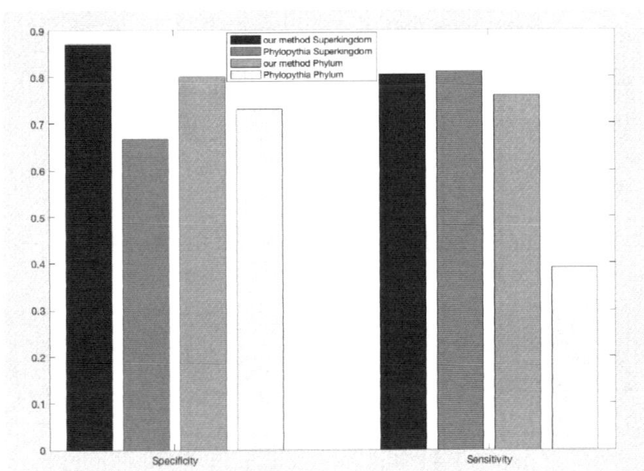

Figure 2. Comparison of Specificity and Sensitivity of the proposed method versus popular existing methods at superkingdom and Phylum levels using widely used simHC data set.

3.2 Horizontal Gene Transfer

Horizontal gene transfer (HGT) is the transfer of genetic materials between organisms other than by the transmission of it from parent to offspring. HGT is commonly observed among prokaryotes (e.g. from archea to bacteria). Obviously, this phenomenon adds another layer of complexity to the OTU assignment of metagenomic fragments and would confused the classifier in making the decision. As a result, the classification results can in turn reflect the potential underlying HGT events among prokaryotes. We observe that relatively highly misclassification among certain organisms, such as Escherichia coli, Bacillus subtilis, and Methanobacterium thermoautotrophicum, indicating the potential HGT event among these organism, as supported from literature [35].

4 Conclusions

Metagenomics has gained tremendous attention with the advance of computer engineering and bioinformatics [1][36]. In this paper, we have investigated a new approach for metagenomic OTU assignment using PCA-initialized deep learning using Relu activation. We demonstrated that this proposed method is very efficient to tackle the problem of metagenomics OTU assignment.

5 References

[1] Nageswara Rao Reddy Neelapu1 and Challa Surekha. "Next-Generation Sequencing and Metagenomics". researchgate, 2015.

[2] Zheng, Hao, and Hongwei Wu. "Short prokaryotic DNA fragment binning using a hierarchical classifier based on linear discriminant analysis and principal component analysis." Journal of bioinformatics and computational biology 8.06 (2010): 995-1011.

[3] Nakamura, K., T. Oshima, T. Morimoto, S. Ikeda, H. Yoshikawa, Y. Shiwa, S. Ishikawa, M.C. Linak, and A. Hirai, et al. 2011. Sequence-specific error profile of Illumina sequencers. Nucleic Acids Res. 39 (13): e90.

[4] Hess, M., A. Sczyrba, R. Egan, T.W. Kim, H. Chokhawala, and G. Schroth. 2011. Metagenomic discovery of biomass degrading genes and genomes from cow rumen. Science. 331 (6016): 463-67

[5] Kanj, Sawsan, Thomas Bruls, and Stephane Gazut. "Shared Nearest Neighbor clustering in a Locality Sensitive Hashing framework." *bioRxiv* (2016): 093898.

[6] Lee, Jongin, et al. "FCMM: A comparative metagenomic approach for functional characterization of multiple metagenome samples." *Journal of microbiological methods* 115 (2015): 121-128.

[7] Delmont, Tom O., et al. "Structure, fluctuation and magnitude of a natural grassland soil metagenome." *The ISME journal* 6.9 (2012): 1677-1687.

[8] Aagaard, Kjersti, et al. "A metagenomic approach to characterization of the vaginal microbiome signature in pregnancy." PloS one 7.6 (2012): e36466.

[9] Hasan, Nur A., et al. "Microbial community profiling of human saliva using shotgun metagenomic sequencing." PLoS One 9.5 (2014): e97699.

[10] Venter, J. Craig, et al. "Environmental genome shotgun sequencing of the Sargasso Sea." *science* 304.5667 (2004): 66-74.

[11] Biers, Erin J., Shulei Sun, and Erinn C. Howard. "Prokaryotic genomes and diversity in surface ocean waters: interrogating the global ocean sampling metagenome." *Applied and environmental microbiology* 75.7 (2009): 2221-2229.

[12] Rusch, Douglas B., et al. "The Sorcerer II global ocean sampling expedition: northwest Atlantic through eastern tropical Pacific." *PLoS Biol* 5.3 (2007): e77.

[13] Kurokawa, Ken, et al. "Comparative metagenomics revealed commonly enriched gene sets in human gut microbiomes." *Dna Research* 14.4 (2007): 169-181.

[14] Fierer, Noah, et al. "The influence of sex, handedness, and washing on the diversity of hand surface bacteria." *Proceedings of the National Academy of Sciences* 105.46 (2008): 17994-17999.

[15] Mathieu, Alban, et al. "Life on human surfaces: skin metagenomics." *PLoS One* 8.6 (2013): e65288.

[16] Zheng, Hao, and Hongwei Wu. "A novel LDA and PCA-based hierarchical scheme for metagenomic fragment binning." *Computational Intelligence in Bioinformatics and Computational Biology, 2009. CIBCB'09. IEEE Symposium on.* IEEE, 2009.

[17] Zheng, Hao, et al. "CpGIMethPred: computational model for predicting methylation status of CpG islands in human genome." *BMC medical genomics* 6.1 (2013): S13.

[18] Huson, D. H., Auch, A. F., Qi, J., and Schuster, S. C. (2007). MEGAN analysis of metagenomic data. Genome Research, 17(3), 377–386.

[19] McHardy, Alice Carolyn, et al. "Accurate phylogenetic classification of variable-length DNA fragments." Nature methods 4.1 (2007): 63-72.

[20] Rosen, G. L., Reichenberger, E. R., and Rosenfeld, A. M. (2011). NBC: the naive bayes classification tool webserver for taxonomic classification of metagenomic reads. Bioinformatics, 27(1), 127–129.

[21] Brady, A. and Salzberg, S. L. (2009). Phymm and PhymmBL: metagenomic phylogenetic classification with interpolated markov models. Nature Methods, 6(9), 673–676.

[22] Mohammed, M. H., Ghosh, T. S., Singh, N. K., and Mande, S. S. (2011). SPHINX¡aan algorithm for taxonomic binning of metagenomic sequences. Bioinformatics, 27(1), 22–30.

[23] Albertsen, M., Hugenholtz, P., Skarshewski, A., Nielsen, K. L., Tyson, G. W., and Nielsen, P. H. (2013). Genome sequences of rare, uncultured bacteria obtained by differential coverage binning of multiple metagenomes. Nature Biotechnology, 31(6), 533–538.

[24] Baran, Y. and Halperin, E. (2012). Joint analysis of multiple metagenomic samples. PLoS Computational Biology, 8(2), e1002373.

[25] Yang Young Lu, Ting Chen, Jed A. Fuhrman, Fengzhu Sun; COCACOLA: binning metagenomic contigs using sequence COmposition, read CoverAge, CO-alignment and paired-end read LinkAge. *Bioinformatics* 2017; 33 (6): 791-798.

[26] http://fames.jgi-psf.orgcgi-bindataset desc.pl?dataset=all

[27] Strous, Marc, et al. "The binning of metagenomic contigs for microbial physiology of mixed cultures." Frontiers in microbiology 3 (2012): 410.

[28] Bentley, Stephen D., and Julian Parkhill. "Comparative genomic structure of prokaryotes." Annu. Rev. Genet. 38 (2004): 771-791.

[29] Teeling H, Waldmann J, Lombardot T, et al. . TETRA: a web-service and a stand-alone program for the analysis and comparison of tetranucleotide usage patterns in DNA sequences, BMC Bioinformatics , 2004, vol. 5 pg. 163

[30] Chan CK, Hsu AL, Halgamuge SK, et al. . Binning sequences using very sparse labels within a metagenome, BMC Bioinformatics , 2008, vol. 9 pg. 215

[31] Chatterji S, Yamazaki I, Bai Z, et al. . CompostBin: a DNA composition-based algorithm for binning environmental shotgun reads, Res in Comp Mol Biol (LNCS) , 2008, vol. 4955 (pg. 17-28)

[32] Wu YW, Ye Y. A novel abundance-based algorithm for binning metagenomic sequences using l-tuples, J Comput Biol , 2011, vol. 18 3(pg. 523-34)

[33] Leung HCM, Yiu SM, Yang B, et al. . A robust and accurate binning algorithm for metagenomic sequences with arbitrary species abundance ratio, Bioinformatics , 2011, vol. 27 11(pg. 1489-95)

[34] Seuret, Mathias, et al. "PCA-Initialized Deep Neural Networks Applied To Document Image Analysis." arXiv preprint arXiv:1702.00177 (2017)

[35] Garcia-Vallvé, Santiago, Anton Romeu, and Jaume Palau. "Horizontal gene transfer in bacterial and archaeal complete genomes." *Genome Research* 10.11 (2000): 1719-1725.

[36] Taizhi Liu, Chang-Chih Chen and Linda Milor, "Comprehensive Reliability-Aware Statistical Timing Analysis Using a Unified Gate-Delay Model for Microprocessors," IEEE Transactions on Emerging Topics in Computing, 2016

An in-silico Construction of Plausible trans- and cis-Elements of Non-housekeeping Genes

Edward A. Salinas[1], **Amitava Karmaker**[2*(corresponding author)]
[1]Independent Researcher, Cambridge, MA, 02140, USA
[2]Univ. of Wisconsin-Stout, Menomonie, WI, 54751

Abstract - *Regulation through transcription factors is an important attribute to the complexity of an organism. Determining the transcription factors that regulate a gene in different cell types will help lay the foundation for building gene regulatory networks to understand their complexity. More importantly, it may lead to possible treatment of complex diseases, such as cancer, through transcriptional control. Here we present an exploratory analysis of expression data to better understand relationships among genes and transcription factors and identify possible genomic binding sites where physical interactions may occur and perhaps be validated.*

Keywords: transcription-regulatory-networks, hierarchical clustering, correlation coefficients, cancer, motifs, transcription-factor binding sites

1 Introduction

Comparative genomic analyses suggest that the number of genes among eukaryotic cells does not vary widely, and thus cannot account for the complexity of an organism. For examples, the simple nematode *C. elegans* possess about 20,000 genes while the number of human genes is estimated to be between 20,000 and 25,000[16, 17]. The complexity of an organism is thought to be attributed in part to elaborate transcriptional control mechanism via cis-element regulation and alternative splicing [18, 19]. Furthermore, many diseases have been associated with irregularities in transcriptional control process. Genome-scale investigations of the transcriptional process will not only help to shed light on the mystery of how highly-evolved organisms achieve their complexity through so few genes, but more importantly perhaps lead to treatments of complex diseases (like cancer) through transcriptional control[20].

To regulate transcription and to control the expression of genes, transcription factors bind to DNA at specific promoter or enhancer sites and thus either facilitate or inhibit the gene expression. When present at sufficient concentration and properly activated, transcription factor proteins bind to transcription factor binding sites (TFBS). Thus, coding regions of the genome are transcribed to mRNA and become available for translation. Correlation studies between transcription factors and corresponding genes and controlling elements allow us to construct the basic components of gene regulatory networks[21].

Microarray technology has been extensively used to monitor gene expression in studies of transcriptional regulators and/or disease progression [21,22]. Sets of genes correlated with tumor type and recovery prognosis in colorectal cancer has been previously found using expression arrays [22].

In order to better understand gene regulation, we performed exploratory data analyses focusing on which transcription factors may regulate a particular gene. We analyzed microarray expression data that is a survey of genetic expression. Exploring further, we also located conserved regions around the promoter regions that have the potential of being the binding sites of these transcription factors. Our findings demonstrate some approaches for building gene regulatory networks

2 Materials and Methods

We downloaded spotted cDNA microarray gene expression data of normal human tissues obtained from a data repository [1]. The data provide us with 26,260 unique genes 35 different organs to analyze. In total, the data set consists of 115 tissue specimens. For each experimental tissue sample, Cy5- and Cy3- labeled samples were co-hybridized to a cDNA microarray containing 39,711 human cDNA's, representing 26,260 different genes. The data came from a normal human female and pooled cell lines. Expression ratios were globally normalized by mean-centering each gene across all arrays.

An initial exploration of the data consisted of a two-way hierarchical clustering procedure. From that we created a dendrogram that clusters co-expressed tissue-specific genes At each leaf node of this dendrogram, we performed t-tests (Eq 1) to compare the expressions of samples present at the current node against the expressions of the rest of the samples with a 95% confidence interval ($\alpha = 0.05$). This identified a pool of genes (including transcription factor genes) that have low variances among their expression levels, but have significant variances over peer group of genes at other branches. (See Figure 1).

$$t = \frac{(\overline{X}_1 - \overline{X}_2)}{S_{\overline{X}_1-\overline{X}_2}}, \text{ where } S_{\overline{X}_1-\overline{X}_2} = \sqrt{\frac{S_1^2}{n_1} + \frac{S_2^2}{n_2}} \quad \text{Eq (1)}$$

$\overline{X}_i, S_i^2, n_i$ mean, variance and population size

In addition, We used Pearson's correlation coefficient, as shown in equation 2, to identify pairs of gene and TF among the genes in each node that are highly correlated. The values of Pearson's correlation coefficient range from -1 to +1. Any value in positive scale indicates increasing linear relationship with +1 being perfectly linear correlated and negative values denote the case of a negative linear relationship. Any value in between in all other cases represents the degree of linear dependence between the variables (i.e. gene and TF pair). The correlation coefficients provide indication showing how genes are up-regulated and down-regulated with respect to transcription factors.

$$r_{xy} = \frac{n\sum x_i y_i - \sum x_i y_i}{\sqrt{n\sum x_i^2 - (\sum x_i)^2}\sqrt{n\sum y_i^2 - (\sum y_i)^2}} \quad \text{Eq (2)}$$

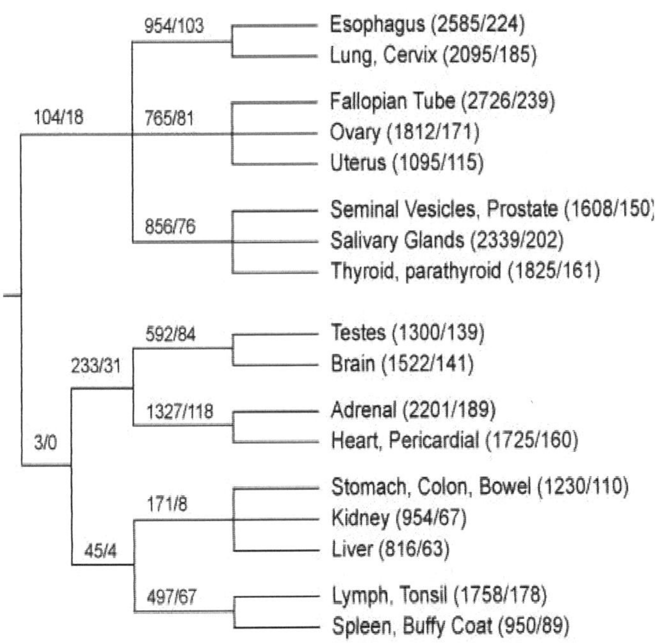

Fig 1. Illustration of t-test approach on the hierarchical tree structure of the available samples of Microarray. Each node is associated with two numbers a/b, where a and b are the number of genes and TF's associated with the group of tissues represented by that node. For example, we found 954 genes and 103 transcription factors that are common in Esophagus, Lungs and Cervix.

3 Analysis and Discussion

We first performed hierarchical clustering on the data set to group the tissues according to the gene expression profile, and determined the subsets of genes that are co-expressed in each of these groups. The clustering result is shown in the tree in Figure 1. Second, we performed t-test (see Methods) at every node in a bottom-up fashion to identify sets of genes (including transcription factor encoding genes) that have low variance among themselves, but vary significantly than those in the peer branches. We applied Pearson's Correlation Co-efficient to construct pairs of gene and transcription factor in each of these gene sets that are highly correlated. Table 1 and 2 show the top ten most positively and negatively correlated genes and transcription factors.

Transcription Factor	Gene	Correlation Co-efficient
TAF9L	SPIN	0.7644
TAF9L	CPSF5	0.7611
MLLT10	RYK	0.7535
PIAS1	SRPK2	0.7343
ZNF83	FLJ10618	0.7317
ZNF83	SPIN	0.7305
PIAS1	CD47	0.7263
RBM9	DSTN	0.7151
ZNF83	WSB1	0.7139
ELF1	ZNF217	0.7129
...	...	...
TAF9L	DPP7	-0.6186
ELF1	CRTAP	-0.6096
TAF9L	DPP7	-0.6046
TAF9L	INHBB	-0.5595
ELF1	ALS2CR3	-0.5549
MLLT10	DPP7	-0.5537
PIAS1	PLG	-0.5459
MAF	DPP7	-0.5442
ELF1	MYO10	-0.5439
MXI1	GABBR1	-0.5431

Table 1: list of top 10 most positively/negatively correlated genes with corresponding transcription factors

3.1 Oncogenes

We identified some cancer genes linked to AML (Acute Myeloid Leukemia) and CML (Chronic Myelogenous Leukemia) to explore consistencies with known interactions. As discussed below, some highly correlated TF-gene pairs have been shown to bear regulatory relationship. This serves to provide some evidence to the validity of the results.

One of the well-recognized oncogenes for AML (Acute Myelogenous Leukemia) is MLL (myeloid/lymphoid or mixed-lineage leukaemia, Loc. 11q23) [5, 6]. In our analysis, we found MDM2 (transformed 3T3 cell double minute 2, p53 binding protein, Loc. 12q14.3-q15), which is a potential target of tumor suppressor protein p53, to be up-regulated with MLL. Over-expression of MDM2 may cause excessive inactivation of tumor protein p53, deteriorating its tumor suppressor function [7, 8].

Another interesting find was for MYC (v-myc myelocytomatosis viral oncogene homolog, Loc. 8q24.12-q24.13) which is overexpressed in cases of ALL (Acute Lymphocytic Leukaemia). Our results suggest a correlation of MYC with KLF4 (Kruppel-like factor 4, Loc. 9q31), a Kruppel-like factor believed to regulate neurogenesis and cell cycle progression [9] and JUNB (Jun B proto-oncogene, Loc. 19p13.2), which elevates the transcription of v-src via transmission of mitogenic signal [10].

The DEK (DEK oncogene, Loc. 6p23) gene produces a fusion with the CAN protein (9q34) in a subtype of AML patients [11, 23]. We found HMGB1 (high-mobility group box 1, Loc. 13q12) in our analysis, and it was also reported in the literature as a gene which controls the binding behavior of DEK [12]. Another transcription factor correlated with DEK is sp3 (Sp3 transcription factor, Loc. 2q31) [13].

Our analyses on Oncogenes provide some supporting evidences for the correlations between transcription factors and corresponding genes. Most often, the findings are consistent with what we found in biological literature surveys. Also we find a number of significant candidates that may be quite relevant to cancer studies. For example, NR2F6, which is a nuclear receptor, is positive correlated with breast cancer genes (ErbB2, CCND1) and one of the Leukemia genes (BCR), while it is inversely correlated with some other Leukemia genes (MYC, DEK).

3.2 Searching for possible cis-elements and Transcription-factor Binding Sites

We searched for the consensus sequences (10, 9, 8-mers) among the three Leukaemia genes namely BCR, DEK and MYC. The sequences were collected from the promoter regions of the respective genes 700 bp upstream. In order to discover consensus substring of DNA sequences, we ran a brute-force routine so that n-mers of one gene matches with another. Here n were set to 10, 9 and 8 to obtain sequences as long as possible. The resulting n-mers are given in table 2. There were no sequences longer than 10-mers, and obviously most of sequences were found for 8-mers. The table also contains some other information, such as epd ratio (frequency of the sequences in the Eukaryotic Promoter Database [14] in respect of all sequences, expressed in thousands) and likewise refseq ratio (NCBI Reference Sequences [15]). As usual, the sequences containing GC-box received higher ratio counts.

To identify prospective TFBS (cis-elements), we concentrate more on the n-mers that appear multiple times across the promoter sequences. We ignored the possible GC-boxes from the short-listed candidates. Then we analyzed the promoter sequences of all the genes, including BCR, DEK and MYC, which are co-regulated by same transcription factors. We tried to search for cis elements that appear within multiple genes and that have relatively low ratio counts. These finding are also shown in the table with some putative examples highlighted as well as in a schematic figure.

Consensus sequence of *cis*-elements	Occurrence in Gene	EPD ratio	Refseq ratio
10-mers			
CCTTCCTGCG	**B, D**	**3.53535**	**10.60606**
GAGGCGCCCT	M, D	3.68324	23.94107
9-mers			
AGGCGCCCT	**M, D**	**2.47191**	**16.40449**
CCCGCCCTG	B, D	2.49465	15.75196
CCTTCCTGC	**B, D**	**0.32751**	**2.77419**
CCTGGCTCC	B, M	0.71168	3.28909
GAGGCGCCC	M, D	1.62303	14.17442
GCCCTCCGC	B, D	7.11013	28.12451
8-mers			
CCCTCCCC	**B, M**	**1.00492**	**6.65394**
CCCGCCCT	**B, D**	**4.34308**	**21.68874**
CCCCGGCC	B, M	4.21206	25.83113
GAGCCGGC	**B, D**	**2.88072**	**23.3387**
GCAGAGGG	**B, M**	**0.57759**	**3.15502**
GGGCCTCA	**B, D**	**0.45299**	**2.73807**
TACGCGCG	**M, D**	**13.49693**	**51.53374**
CCCGCCCT	**B, D**	**4.34308**	**21.68874**

Table 2: Potential cis-elements for BCR (B), DEK (D) and MYC (M) with their frequency information from two repositories.

4 Conclusions

In this study, we have constructed for each gene in our data set, a list of prospective transcription factors that may regulate it. For those genes that seem to share a significant set of common transcription factors, we also constructed a list of putative cis-elements. We selected a small set of Leukaemia genes, among 26,260 genes, to illustrate the potential of using our approach to identify transcription factors.

5 References

[1] Shyamsundar R., Kim YH., Higgins JP, Montgomery K, Jorden M, Sethuraman A., Van de Rijn M., Botstein D, Brown PO, Pollack JRA. DNA microarray survey of gene expression in normal human tissues. Genome Biol. 2005; 6(3): R22.

[2] Kurzrock R, Kantarjian HM, Druker BJ, Talpaz M. Phildelphia Chromosome-positve leukemias: from basic mechanisms to molecular therapeutics. Ann Intern Med (2003), 138:819–30.

[3] Stanulla M, Schünemann HJ, Thandla S, Brecher ML, Aplan PD. Mol Pathol. Pseudo-rearrangement of the MLL gene at chromosome 11q23: a cautionary note on genotype analysis of leukaemia patients. (1998) Apr; 51(2): 85-89.

[4] Deng LW, Chiu I, Strominger JL. MLL 5 protein forms intranuclear foci, and overexpression inhibits cell cycle progression. Proc Natl Acad Sci U S A. (2004), Jan 20; 101(3): 757-762.

[5] Puente XS, Velasco G, Gutiérrez-Fernández A, Bertranpetit J, King MC, López-Otín C. Comparative analysis of cancer genes in the human and chimpanzee genomes. BMC Genomics. (2006), 7: 15.

[6] Yamamoto S, Nishi M, Taniguchi K, Imayoshi M, Ogata Y, Iwanaga M, Sakai N, Hamasaki Y, Ishii E. Partial tandem duplication of MLL gene in acute myeloid leukemia with translocation (11;17)(q23;q12-21). Am J Hematol. (2005), Sep; 80(1):46-9.

[7] Best JL, Amezcua CA, Mayr B, Flechner L, Murawsky CM, Emerson B, Zor T, Gardner KH, Montminy M. Identification of small-molecule antagonists that inhibit an activator:coactivator interaction. Proc Natl Acad Sci U S A. (2004), Dec 21; 101(51): 17622-17627.

[8] Wiederschain D, Kawai H, Gu J, Shilatifard A, Yuan ZM. Molecular Basis of p53 Functional Inactivation by the Leukemic Protein MLL-ELL. Mol Cell Biol. 2003 Jun; 23(12): 4230-4246.

[9] Smaldone S, Laub F, Else C, Dragomir C, Ramirez F. Identification of MoKA, a Novel F-Box Protein That Modulates Krüppel-Like Transcription Factor 7 activity. Mol Cell Biol. 2004 Feb; 24(3): 1058-1069.

[10] Apel I, Yu CL, Wang T, Dobry C, Van Antwerp ME, Jove R, Prochownik EV. Regulation of the junB gene by v-src. Mol Cell Biol. 1992 Aug; 12(8): 3356-3364.

[11] Waldmann T, Scholten I, Kappes F, Hu HG, Knippers R. The DEK protein--an abundant and ubiquitous constituent of mammalian chromatin. Gene. 2004 Dec 8; 343(1):1-9. Review.

[12] Waldmann T, Baack M, Richter N, Gruss C. Structure-specific binding of the proto-oncogene protein DEK to DNA. Nucleic Acids Res. 2003 Dec 1; 31 (23): 7003-10.

[13] Lin DY, Fang HI, Ma AH, Huang YS, Pu YS, Jenster G, Kung HJ, Shih HM. Negative modulation of androgen receptor transcriptional activity by Daxx. Mol Cell Biol. 2004 Dec; 24(24):10529-41.

[14] The Eukaryotic Promoter Database 87 [http://www.epd.isb-sib.ch/]

[15] NCBI Reference Sequence (RefSeq)[www.ncbi.nlm.nih.gov/RefSeq/]

[16] I. Ezkurdia, et al, Multiple evidence strands suggest that there may be as few as 19,000 human protein-coding genes Human Mol. Genet (2014), 23(22):5866-78

[17] Hillier LW, et al., Genomics in C. elegans: so many genes, such a little worm, Genome Res. 2005 Dec;15(12):1651-60

[18] De Mendoza A, Sebé-Pedrós A, Šestak MS, et al. Transcription factor evolution in eukaryotes and the assembly of the regulatory toolkit in multicellular lineages. , PNAS 2013;110(50):E4858-E4866. doi:10.1073/pnas.1311818110.

[19] Lu Chen, Stephen J. Bush, et al.,; Correcting for Differential Transcript Coverage Reveals a Strong Relationship between Alternative Splicing and Organism Complexity. Mol Biol Evol 2014; 31 (6): 1402-1413

[20] Dominik E. Dorer, Dirk M. Nettelbeck, Targeting cancer by transcriptional control in cancer gene therapy and viral oncolysis, Advanced Drug Delivery Reviews Volume 61, Issues 7–8, 2 July 2009, Pages 554–571

[21] Zhi-Ping Liu, Reverse Engineering of Genome-wide Gene Regulatory Networks from Gene Expression Data, Curr Genomics. 2015 Feb; 16(1): 3–22.

[22] Nurul Ainin Abdul Aziz, A 19-Gene expression signature as a predictor of survival in colorectal cancer, BMC Medical Genomics , 2016 9:58

[23] Zhou MH, Yang QM. NUP214 fusion genes in acute leukemia (Review). Oncology Letters. 2014;8(3):959-962. doi:10.3892/ol.2014.2263.

A comparison of methods for classifying promoter regions in E. coli based on structural properties of DNA

C. Wright[1], **J. Kaur**[2], **A. S. Newsome**[2], **and C. Bland**[2]
[1]Department of Mathematics, Jackson State University, Jackson, MS, USA
[2]Bioinformatics Program, Mississippi Valley State University, Itta Bena, MS, USA

Abstract - *One of the major challenges in biology is the correct identification of promoter regions. Computational methods based on motif searching have been the traditional approach taken. Studies have shown that DNA structural properties, such as free energy, curvature, and stress-induced duplex destabilization (SIDD) are useful in promoter classification, as well. In this paper, these properties were compared for their effectiveness in correctly classifying promoters. When using a decision tree for promoter classification based on DNA structural properties, SIDD showed a slight improvement over free energy and curvature, with f-score values 70.9%, 67.1%, and 61.5%, respectively.*

Keywords: promoter classification, DNA curvature, SIDD, free energy

1 Introduction

Identification of promoters is an important issue in biology, given that they are central in understanding the process by which genes are regulated. Wet-lab methods for promoter identification provide accuracy but suffer from being time-consuming. To facilitate faster processing, computational methods are required. Although far from perfect, they do provide a means for quickly identifying potential targets for experimental validation.

Several computational methods for promoter classification have been proposed. Most include some analysis of sequence patterns commonly found in promoter regions, such as -10 and -35 motifs [1, 2]. However, these patterns are not always sufficiently conserved to allow for adequate classification. Furthermore, there are clearly other factors not directly related to sequence motifs that are closely associated with promoter regions.

Promoter regions have unique characteristics in their physical structure that play major roles in transcription by facilitating protein-DNA interactions. Some of these properties include GC skew, bendability, free energy, curvature, base stacking, and stress-induced duplex destabilization (SIDD). Studies have reported impressive results using DNA structural properties for identifying promoter regions [3, 4, 5, 6]. This study assesses the feasibility of a computer-based classification approach for promoter identification in prokaryotes based on DNA free energy [7], curvature [8], and SIDD [9].

2 Methods

Analysis was performed on the genome of *E. coli K12*. Each sequence value was converted to its corresponding numeric structural property value.

2.1 Dataset

The whole genome of *E. coli K12* was downloaded from NCBI. Experimentally verified transcription start sites were obtained from the Regulon database (Release: 6.4) [10]. This database release provided a compilation of 1771 promoter sequences. The dataset was filtered for unique promoters with known TSS locations, resulting in 1648 records.

Structural profiles were computed from the sequence data. The SIDD profile computations were obtained from Benham [5]. The free energy profile was computed using the nearest-neighbor thermodynamic parameters of base pairings described in [11]. The curvature profile was computed using the CURVATURE program [12, 13].

2.2 Classification

The training and testing datasets were constructed from the *E. coli K12* structural profile data. Positive instances (promoters) were defined as the 500 bp region from -400 to +100, with respect to TSSs. This dataset was composed of 1648 positive instances and 4944 negative instances, which represents a 3:1 ratio of negatives and positives. A randomly selected two-third and one-third split was used for training and testing data, respectively. The Weka data mining suite [14] was used to perform the classifications using its J48 decision tree.

2.3 Evaluation Measures

Classification results were used to evaluate the predictability of the structural properties. In order to compare predictions using a one-dimensional performance measure, the weighted average of the precision and recall (known as f-score) was computed for curvature, free energy and SIDD.

Precision, recall, and f-score were defined as follows,

$$precision = \frac{TP}{TP + FP} \quad (1)$$

$$recall = \frac{TP}{TP + FN} \quad (2)$$

$$f\text{-}score = \frac{2 \text{ x } precision \text{ x } recall}{precision + recall} \quad (3)$$

where *TP*, *TN*, *FP*, and *FN* are the numbers of true positives, true negatives, false positives and false negatives, respectively.

3 Results and discussion

A comparison of free energy, curvature, and SIDD structural profiles is shown in the following figures. To create the structural data, each sequence value in *E. coli K12* was converted to its corresponding numeric structural property value. Next, the average value at each location was computed for all promoters (for the 500 bp region from -400 to +100, with respect to transcription start sites at +1).

3.1 Signatures of structural properties

Figure 1 is DNA free energy. High free energy values indicate low stability, and indicate regions where strand separation is more likely to occur. Figure 1 shows a low stability region from -100 to +50, with respect to the TSS. A distinctive peak appears near -10. So, the -10 region may be the least stable.

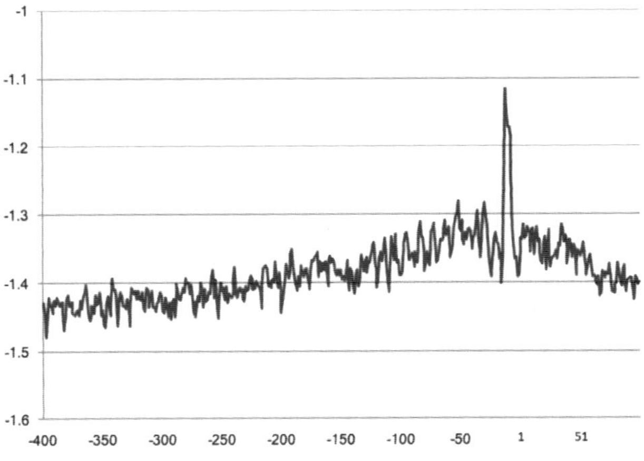

Figure 1: Average free energy values for the promoter regions

Similar changes in promoter regions can be seen in Figure 2 for SIDD, represented as *G*(*x*). *G*(*x*) corresponds to the incremental free energy needed for the base pair at position *x* to always remain open. It begins a noticeable decrease until its lowest points near -35 and -10, then begins an increase.

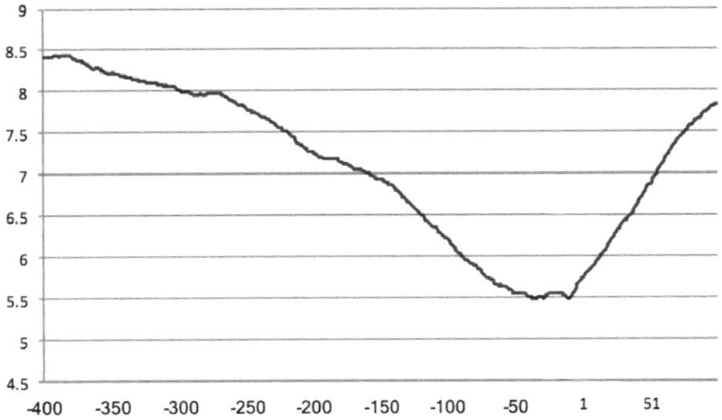

Figure 2: Average SIDD G(x) values for the promoter regions

Curvature increases from -400 to its highest at -53, before beginning to decrease. All three properties show noticeable increases or decreases in promoter regions and distinctive spikes near some known promoter indicators, such as -10, and -35. Thus, structural properties appear to be good candidates for identifying promoter regions.

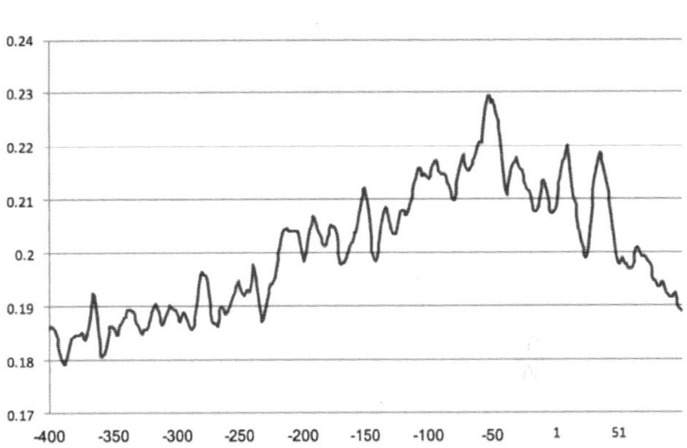

Figure 3: Average DNA curvature values for the promoter regions

3.2 Evaluation

Weka's J48 decision tree was used to perform the classifications of promoters and non-promoters. The construction of the training and testing sets is described in the methods sections. The f-score was computed for curvature, free energy and SIDD. For free energy, the resulting f-score was 67.1% (promoter 50.9%, non-promoter 74.9%); SIDD 70.9% (promoter 56.4%, non-promoter 77.8%); and curvature 61.5% (promoter 42%, non-promoter 71.8%). All methods performed better at identifying non-promoters than promoters. SIDD performed best overall, followed closely by free energy, and then curvature with the lowest f-score.

4 Conclusions

One of the major challenges in biology is the correct identification of promoter regions. Computational methods based on motif searching have been the traditional approach taken. This study has shown that DNA structural properties, such as free energy, curvature, and stress-induced duplex destabilization (SIDD) are useful in promoter classification, as well.

Future research will involve combining multiple structural-based predictors with sequence-based methods. For example, in [5] it was shown that SIDD was not directly related to primary sequences or unique motifs, and not positively correlated with DNA curvature. Thus, using SIDD with other predictive sequence and structural properties, particularly those not strongly correlated, may be fruitful. In addition, it may be useful to determine whether a classifier trained on one genome predicts well on others. Also, combining multiple classifiers as part of a voting system, such as an ensemble, may prove beneficial.

5 References

[1] Gerald Z. Hertz, Gary D. Stormo. "*Escherichia coli* promoter sequences: analysis and prediction"; Methods in Enzymology, Volume 273, Pages 30-42, 1996.

[2] Araceli M. Huerta, Julio Collado-Vides. "Sigma 70 promoters in *Escherichia coli*: specific transcription in dense regions of overlapping promoter-like signals". Journal of Molecular Biology, Volume 333, Issue 2, Pages 261-278, October 2003.

[3] Czuee Morey, Sushmita Mookherjee, Ganesan Rajasekaran, Manju Bansal. "DNA Free Energy-Based Promoter Prediction and Comparative Analysis of Arabidopsis and Rice Genomes". Plant Physiology, Volume 156, Issue 3, Pages 1300-1315, April 2011.

[4] Charles Bland, Abigail S. Newsome, Aleksandra Markovets. "Promoter prediction in *E. coli* based on SIDD profiles and Artificial Neural Networks". BMC Bioinformatics, Volume 11, Supplement 6, October 2010.

[5] Huiqan Wang, Craig J. Benham. "Promoter prediction and annotation of microbial genomes based on DNA sequence and structural responses to superhelical stress". BMC Bioinformatics, 7: 248, May 2006.

[6] Aditi Kanhere, Manju Bansal. "A novel method for prokaryotic promoter prediction based on DNA stability". BMC Bioinformatics, 6:1, January 2005.

[7] Aditi Kanhere, Manju Bansal. "Structural properties of promoters: similarities and differences between prokaryotes and eukaryotes". Nucleic Acids Research, Volume 33, Issue 10, Pages 3165-3175, June 2005.

[8] Limor Kozobay-Avraham, Sergey Hosid, Alexander Bolshoy. "Involvement of DNA curvature in intergenic regions of prokaryotes". Nucleic Acids Research, Volume 34, Issue 8, Pages 2316–2327, May 2006.

[9] Huiqan Wang, Michiel Noordewier, Craig J. Benham. "Stress-induced DNA duplex destabilization (SIDD) in the E. coli genome: SIDD sites are closely associated with promoters". Genome Research, Volume 14, Issue 8, Pages 1575-1584, August 2004.

[10] RequlonDB [http://regulondb.ccg.unam.mx]

[11] John SantaLucia Jr. "A unified view of polymer, dumbbell, and oligonucleotide DNA nearest-neighbor thermodynamics". Proceedings of the National Academy of Sciences USA, Volume 95, Number 4, Pages 1460-1465, February 1998.

[12] E. S. Shpigelman, E. N. Trifonov, A. Bolshoy. "CURVATURE: software for the analysis of curved DNA". Computer Applications in the Biosciences, Volume 9, Issue 4, Pages 435-440, August 1993.

[13] Curvature [http://www.lfd.uci.edu/~gohlke/dnacurve/]

[14] Weka [http://www.cs.waikato.ac.nz]

Para-Seqs: Parallel pattern match tools – mpiTigrScan, mpiGlimmer, smpTigrScan, smpGlimmer

Abhishek Narain Singh[1,2]

[1]AbioTek Consulting Group - ATCG, VA, USA
[2]Department of Computer Science and Interdisciplinary Center of Bioinformatics, Leipzig University, Leipzig, Germany
Email: abi@bioinf.uni-leipzig.de, abhishek.narain@iitdalumni.com
Web: www.tinyurl.com/abinarain

Abstract

A long underpinning need for pattern in genome such as an open reading frame (ORF) or a gene search tool to meet big data analysis world necessity, through distributed or symmetric computer architecture is being met by the work presented in this article. Parallel tools mpiTIGRscan, smpTigrScan, mpiGlimmer & smpGlimmer are presented here for pattern of gene finding application. These software typically deal with nucleotide sequences such as sets of chromosome or whole genome to identify the putative genes . They achieve linear speed-up by segmenting the input data with adequate importance to the segmenting zone from both computational and molecular-biological point of view. This is accomplished by means of both data parallel and task parallel means of. Segmentation of the nucleotide sequence input permits each processor to do the computation on lesser size of the sequence, eliminating disk I/O. It is not advisable to run a non-parallelized gene finder on large sequences since gene finders generally model the number of genes and exons using exponential distributions, so that for longer sequences there is a loss of exons and genes. Genome segmentation under molecular-biological consideration thus also improves the time and reliability of the gene finder result. Data segmentation does not involve heavy communication demand, and these software tools can thus be used trustfully by the computational biologist and bioinformatics community. Apart from software architecture, we also present a detailed performance analysis for demonstrating scalability.

Introduction

Modern biotechnology, biochemical engineering and biology has by far become a highly interdisciplinary subject, to the extent that the molecular nature of the biochemicals need to be studied by reliable software tools, capable of not just assuring low false-positive and false-negatives but also guaranty near reliable results when the data quantity scales up multi-&-manifolds. Big data serves as oil of current accelerated research, discoveries in scientific and business world alike, and to keep up the pace with which the results needs to be generated, appropriate tool should make one equipped of carrying out one's profession competitively. Looking from a supply-chain perspective, if the DNA sequencing companies are the suppliers to the scientists who are engaged in making discoveries, they tend to provide more in terms of data quantity as they leave the overhead of managing the data to the scientist. Thus the loop of demand and supply is being met at either end from DNA sequence provider as well as the scientists who knows that his capability to deal with big data will give him an edge over the competition. Much of the research has now become focused largely on biomolecules since the revolution in molecular-biology since 1970 and thus comes the importance of consensus pattern with signatures that can be mathematically coded with some common features such as in ORFs or genes, for instance. Genome annotation is one of the primary method practiced for drug targeting. Foreign gene is identified in the host by means of sequence comparison algorithms such as an alignment of DNA or protein sequence tool, and thereby strategies are implemented to knockout the gene or the gene-product or a combination in the pathway. Much of the parasites have not yet their genome annotated experimentally and the gene finders thus play crucial role. The situation is worse for higher organisms, which act as host for these parasites, since the wet-lab techniques have their own resource and time constraint. As we see from the big data genome analysis projects which demands reconstruction of individual genomes in the form of Next Generation Sequencing reads to be assembled and then analyzed for pattern, such as my own work on the Genome of the Netherlands project for an anonymous individual A105, Singh et. al., (2012), it is clear that there is immense need for pattern search tools not just for genes and ORFs sequences but also for any specific format of consensus code such as Sines / Lines and other known and unknown patterns. Another pattern match project that I was personally involved was in tackling the SARS virus epidemic where the need of the hour was to quickly look for all possible patterns in the SARS virus genome for trans-membrane binding domains of various ORFs, calcium ion binding and other co-factor and coenzyme binding pattern in an expressed ORF sequence, intra-cellular and extra-nuclear compartmentalization and localization pattern match consensus pattern sequence, such that once you get these possibilities with various confidence level, you can reconstruct a pathway pattern of how the host and parasite interaction could lead to pyroptosis or apoptosis, Singh et. al., 2004. The pressure of getting a timely and useful results in these cases points to the need for high performance computing in serial pattern match algorithms. Unfortunately, the typical approach to gene prediction have proven to be too slow to keep up along with the present rate of sequence determination, particularly with the eukaryotes where there are more landmarks on the nucleotide sequence to be looked for. We present here open-source SMP (symmetric multi-processor) and MPI (message passing interface) parallel forms of the Glimmer and TigrScan software tools used for prokaryotic and eukaryotic gene scanning respectively, that segments and distributed nucleotide sequence for the processors such that each performs the functions on the assigned data partition generated independently, until there is bottleneck, and starts off with the parallelization again after the bottleneck. Karam et. al.(2009), lists an update and history of symmetric multi-processing (SMP). Gropp et. al. (1996), mentions in his article the portable nature of message passing interface (MPI). GLIMMER (Gene Locator and Interpolated Markov ModelER) was first mentioned about its merit for microbial genome in Salsberg et.al(1998). Majoros W, *et al.* (2004) describes the IMM model built for handling eukaryotic genome for gene search using TigrScan. Subsequent versions of this tool was renamed Genezilla.

One of the advantages of database segmentation is that it reduces the high overhead of disk I/O. The sizes of eukaryotic chromosomes and many prokaryotic genomes are in fact much larger than the core memory on most computers. It is not advisable to run a gene finder on large sequences since gene finders generally model the number of genes and exons using exponential distributions, so that for longer sequences there is a loss of exons and genes. Database segmentation permits each processor to search a smaller portion of the nucleotide sequence , thereby reducing extraneous disk I/O and thus not only reducing the execution time remarkably but also improve reliability of the prediction. Further , sequence segmentation does not produce heavy intercommunication between nodes, thereby facilitating Linear speedup. These software essentially comprises of a script to deal with data distribution by means of fine granularity so that the head node assigns tasks to slave nodes in a round robin fashion, keeping in account of the molecular-biological characteristics of the intersection zone as discussed earlier. Parallelization is then achieved by executing the various functions of the respective software by means of posix-Threads in shared memory architecture typically of symmetric multiprocessor nature and thus the tools are prefixed 'smp', and Message Passing Interface (MPI) for distributed computing architecture thus prefixed 'MPI'. Combination of Perl, C++ and Shell script is used where appropriate. The tools can be executed on clusters with 'schedulers' such as the portable batch submission scripts (PBS), allowing adaptation to resource changes by dynamic re-distribution of split database.

TigrScan and the Algorithm

TigrScan models DNA using a Generalized Hidden Markov Model (GHMM). Alternate parses of DNA (into zero or more gene models) are evaluated under this model. A GHMM is an extension of an HMM in which each state can emit a sequence of symbols at each time unit rather than just a single symbol. Whereas an HMM emits a symbol (stochastically) from the current state and then transitions to another state (also stochastically), a GHMM emits a nonempty string of symbols from the current state before transitioning to the next state. This allows the GHMM to explicitly model the length distributions of gene features (rather than always imposing a geometric distribution as does an HMM), and also permits other forms of dependency modeling not normally feasible with a standard HMM. The output format of TigrScan is GFF.

Glimmer and the Algorithm

Glimmer is a system for finding genes in microbial DNA, especially the genomes of bacteria, archaea, and viruses. Glimmer (Gene Locator and Interpolated Markov Modeler) uses interpolated Markov models (IMMs) to identify the coding regions and distinguish them from noncoding DNA. The IMM approach uses a combination of Markov models from 1st through 8th-order, weighting each model according to its predictive power. The Glimmer system consists of two main programs. The first of these is the training program, build-imm, which builds the Interpolated Markov Modeler. This program takes an input set of sequences and builds and outputs Interpolated Markov Modeler for them. These sequences can be complete genes or just partial ORFs (open reading frames). For a new genome, this training data can consist of those genes with strong database hits as well as very long open reading frames that are statistically almost certain to be genes. The second program is glimmer, which uses this IMM to identify putative genes in an entire genome.

Genome Data Segmentation

The nucleotide input file is split into as many split files as the number of processors specified , considering the molecular-biology of the splitting zone. The splitting zone might actually comprise of a gene and so we need to create the corresponding intersection files . The intersection zone size is guided by the 'gene density' of the organism . The average gene density of microbes is about 1000 base-pairs. In smpGlimmer & mpiGlimmer we have kept the zone to be 14000 base pairs in size to be on the safer side. The sum of the intersection file sizes does not generally exceed more than 4% of the actual input sequence , thereby ensuring not much increase in the net workload. In smpTigrScan & mpiTigrScan the intersection files generated comprise of 80000 base-pairs . This value of 'zonal bases' should be changed as the macro in the corresponding C code before compilation if one wishes to be organism specific and want to cut-down time. 'Table A' provides a general list of some genome sizes and corresponding average gene-density.

Table A

SPECIES	GENOME SIZE(bp)	AVERAGE GENE DENSITY
Homo sapiens	3000 million	1 per 100000 bp
Mus musculus	3000 million	1 per 100000 bp
Drosophila Melanogaster	180 million	1 per 9000 bp
Arabidopsis thaliana	125 million	1 per 4000 bp
Caenorhabditis elegans	97 million	1 per 5000 bp
Saccharomyces cerevisiae	12 million	1 per 2000 bp
Escerichia coli	4.7 million	1 per 1400 bp
H. influenzae	1.8 million	1 per 1000 bp

We see that according to the expectation , as we go to higher and higher organism , gene density decreases in general . This is from the biological fact that the non-coding regions comprise of the regulating and control sites which serve a very important network for the organism for tuning the various biological pathways. In all of the microbial genomes sequenced to date, the average gene length is about the same (1000bp), and genes appear to be similarly spread out along the chromosome in each genome. Hence , we can safely consider the average 2000 bases of the eukaryote *S.cervisiae* as the upper limit for all the prokaryotes for finding a gene. Apart from Gene density, the gene size would be another biological parameter to keep in mind while dealing with genome data granularity generation. In general the prokaryotes have non-weighted median gene size of around 4000 bases and eukaryotes have 22000 bases (Xu et. al. 2006). Figures 1 and 2 gives a graphical distribution of number of species of prokaryotes and eukaryotes vs the coding length observed.

For even safer result we have deliberately considered the intersection zone files to span 14000 bases in the spmGlimmer and mpiGlimmer as we see that 8000 bases would be the typical high end gene size that can be encountered for prokaryotes. The zonal value of 80,000 serves fairly good for most eukaryotes, as we see that 50,000 bases would be the typical high end eukaryotic gene size. The software tool gives flexibility to pass on a revised values of these numbers as arguments on command line should there be a need, such as if interested in querying a pattern gene rich region of the genome such as for higher species such as the human genome or mouse and monkey genomes which are consistently used for experimental verification in a medical laboratory setting. The Glimmer software has the default state of considering the sequence to be that of circular DNA. Care has been take to make the corresponding change such that no split files or intersection files is treated as circular DNA . The last intersection file takes care of the circular behavior of the DNA and care should be taken to stop this file generation in case one wishes to consider non-circular microbial genome.

PARALLEL CONCURRENT WORK-FLOW

The four software essentially all make use of the data segmentation script to start with. Thereafter, the parallelization is achieved by means of round-robin data distribution strategy. These parallel codes in different software tools vary depending on the functions involved in the non-parallel version and the implementation required viz., shared or distributed memory architecture. We have made use of p-Threads (posix-threads) for the former and MPI for the later. The concept and the backbone of the parallel code essentially is more or less the head node distributing task to the worker nodes as soon as it signals itself to be free. Since we know the sequence in which the tasks are done in the non-parallel software, and also know about the dependency and concurrency, jobs can be split to various computing cores for independent functions. Barrier synchronization is done wherever applicable and where there is no choice, as bottlenecks though not desired is still sometimes needed to ensure the efficacy of the result.

Major steps in Parallel Algorithm

The details of concurrency extraction and parallel algorithm is mentioned in the figures 3 & 4 at Supplementary material www.tinyurl.com/abinarain (click on 'Educational Stuffs'). The pseudocode for genome database segmentation is mentioned in Supplementary material as well. The key pain points addressed while developing the scripts are listed below.

- Distribute data files to various processors or use NFS
- Perform independent functions on data files in parallel on the basis of dependency-analysis
- Head node gives data to slave for task operation wherever concurrency is possible
- Where critical, create barrier for synchronization
- Concatenate the generated intermediate result wherever necessary before switching to next operation
- Combine the final Output and remove redundancy

Performance and Analysis

The charts below illustrate for the machine type and the data choice and show the drop in time versus number of computing cores used. In general a high performance for bigger data size such as for eukaryotes was observed. For small data size, the inter-processor communication overhead seemed to effect a lot in terms of speedup.

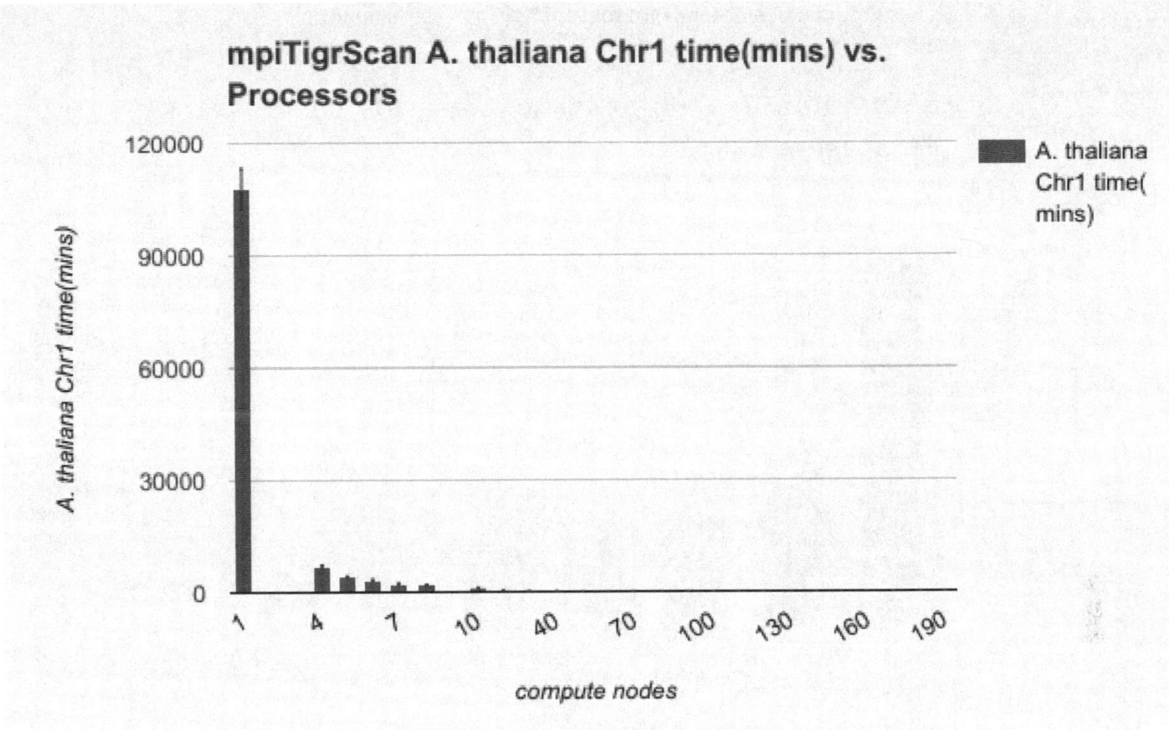

Figure 5: mpiTigrScan performance

The performance in terms of speed-up of mpiTigrScan is so good even at higher number of nodes that a true plot of time vs nodes is not able to make the bars visible to naked eye as we see in figure 5, and thus treatment of y-axis time with natural logarithm creating a semi-log plot makes the bars at higher node visible as in figure 6.

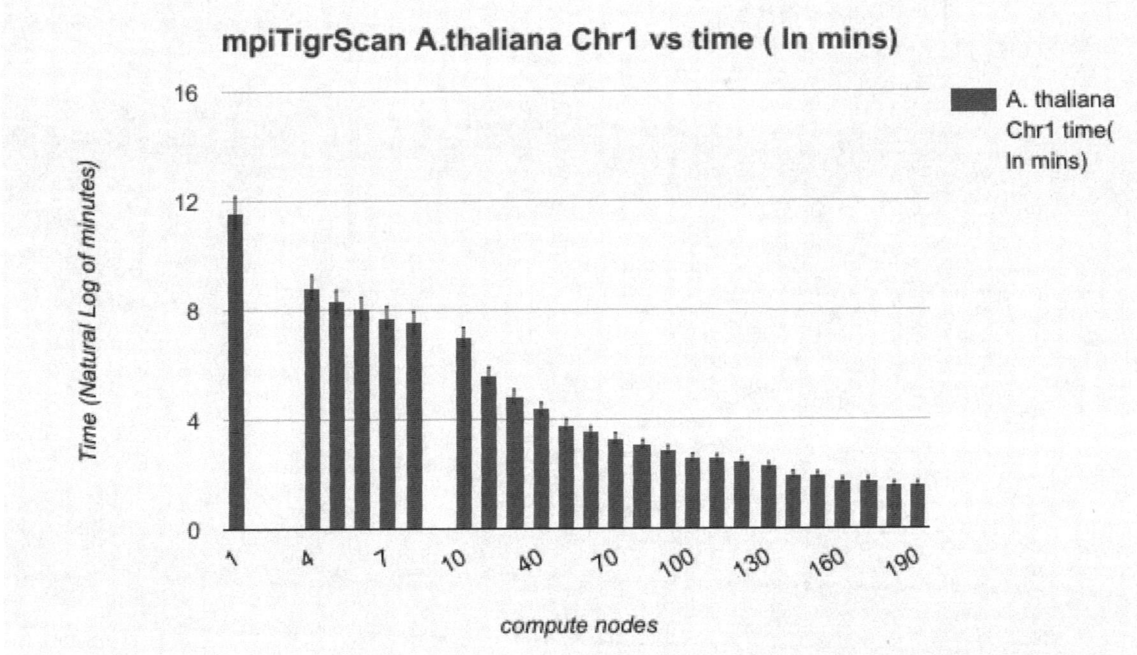

Figure 6: Semi-Log Plot for mpiTigrScan performance

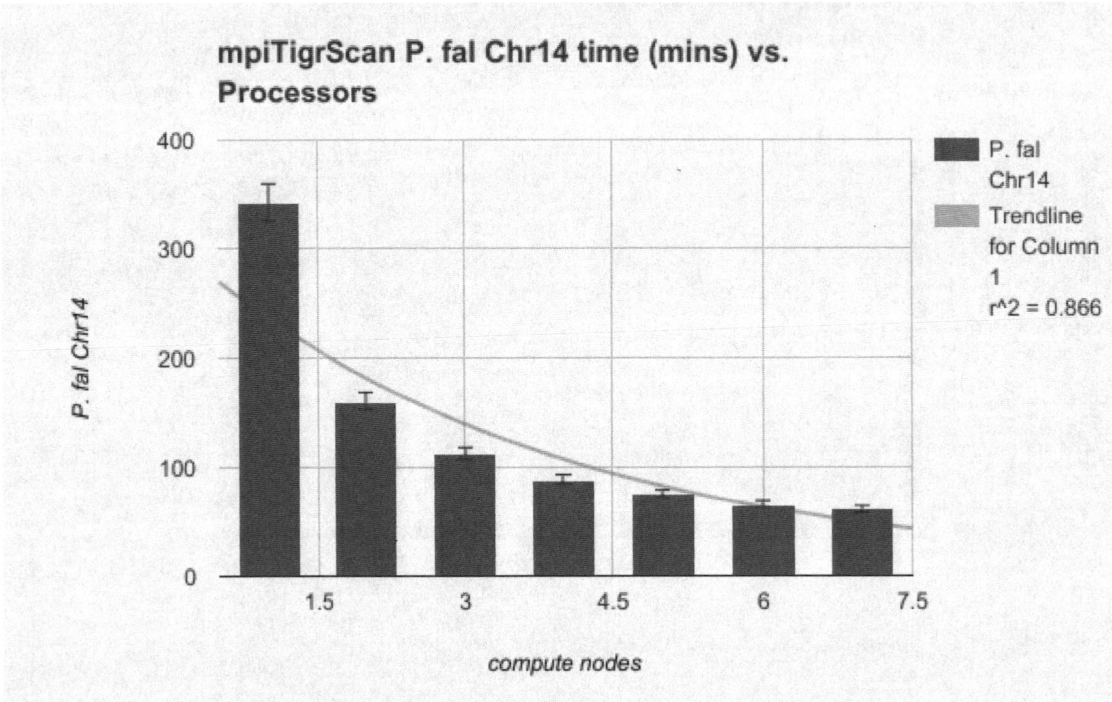

Figure 7: mpiTigrScan performance for *P. falciparum* chromosome 14

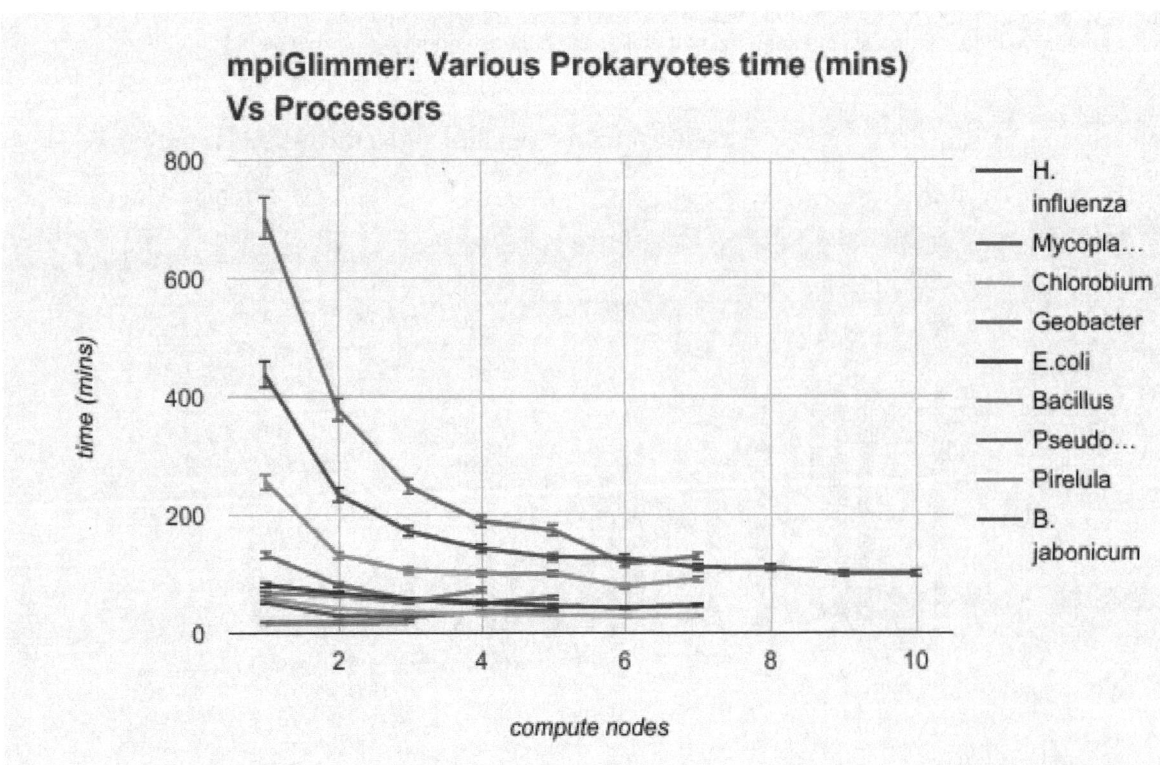

Figure 8: mpiGlimmer perfromance for various prokaryotes

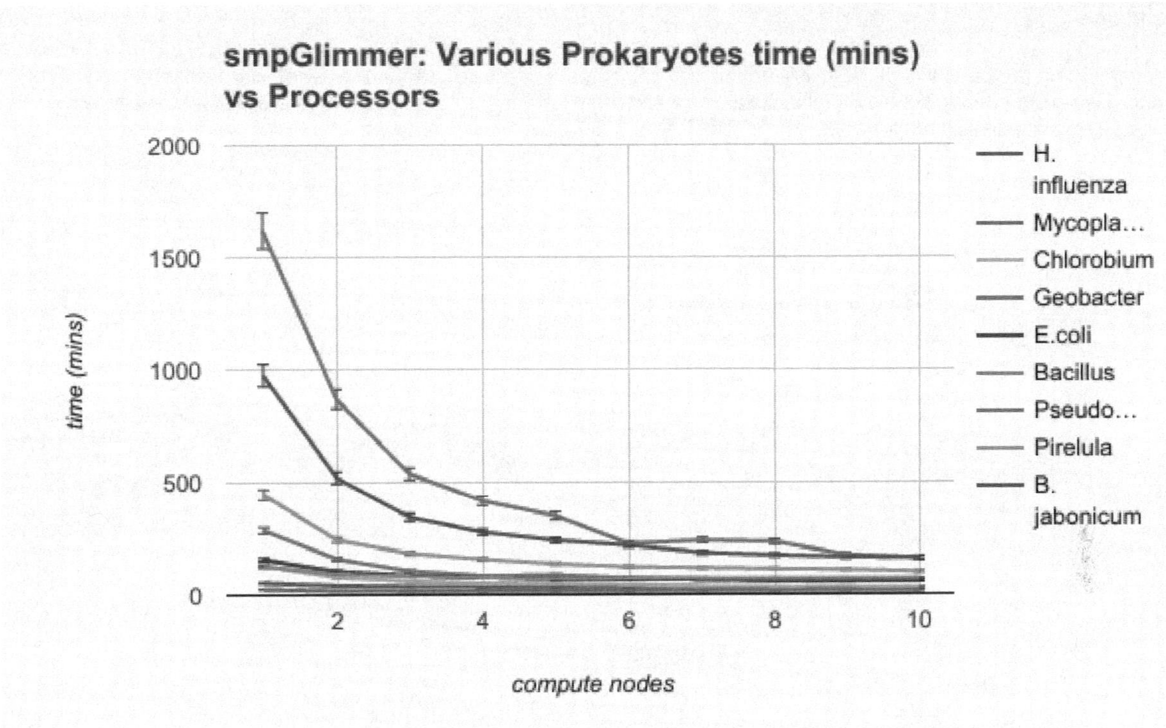

Figure 9: smpGlimmer performance for various prokaryotes

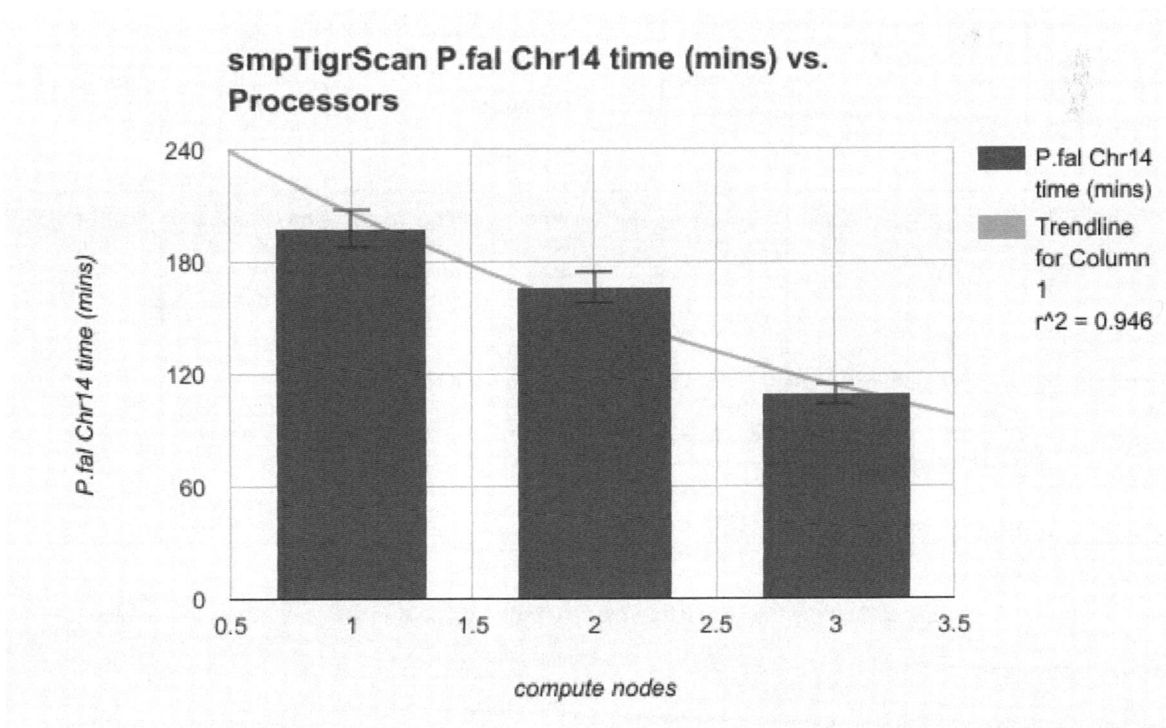

Figure 10: smpTigrScan performance

Benchmarking

Benchmarking the parallel tools with a fixed number of compute cores, 10, against the non-parallel serial version of the tool was done for multiples of similar data sizes and the time it takes for execution. The tools showed linear behavior with increasing multiples of similar data thereby fulfilling the requirements of a robust performance with data and time. To ensure consistency, the same prokaryotic and eukaryotic DNA is used for serial and non-serial versions of the tools.

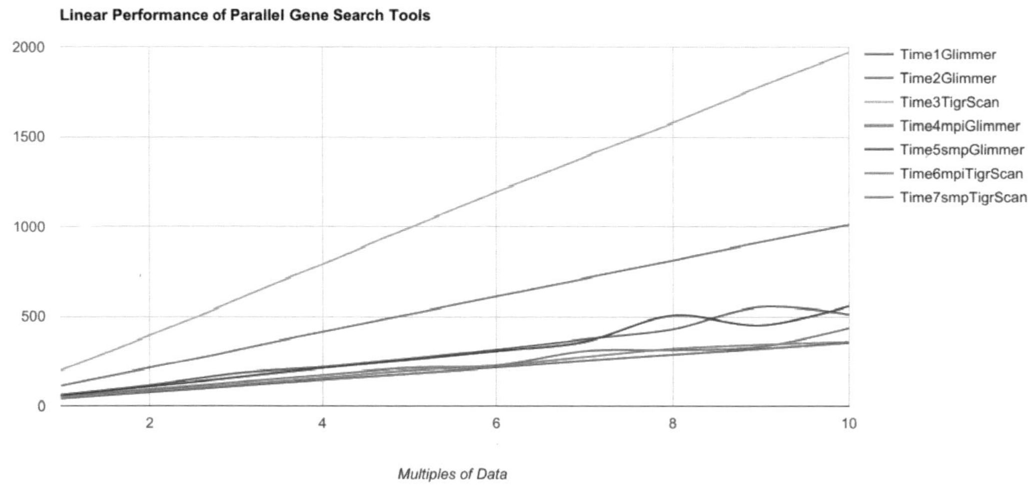

Data Type / Organism	*E.coli*	*E.coli*	Chr 14 *P. falciparum*	*E.coli*	*E.coli*	Chr 14 *P. falciparum*	Chr 14 *P. falciparum*
Tool	Glimmer	Glimmer	TigrScan	mpiGlimmer (with 10 compute cores)	smpGlimmer (with 10 compute cores)	mpiTigrScan(with 10 compute cores)	smpTigrScan(with 10 compute cores)
Machine Used	ece211-10D.eece.unm.edu	linux02.ece.unm.edu	linux03.ece.unm.edu	loslobos.alliance.unm.edu	darwin.cs.unm.edu	loslobos.alliance.unm.edu	linux.ece.unm.edu

Figure 11. Robust time based performance as benchmarked against serial version of the tool

Acknowledgement

The author would like to acknowledge the University of New Mexico Center for Advanced Research Computing technical staffs who were active in the year 2004 and the technical officers engaged in Loslobos computing facilities in the same year for providing their cluster facilities and for providing systems support in general.

Supplementary Materials

Supplementary material at www.tinyurl.com/abinarain . Click on 'Educational Stuffs'.

Funding & Conflict of interest

None to be declared.

References

Abhishek Narain Singh, Dinesh Gupta and Shahid Jameel, Bioinformatics analysis of the SARS virus X1 protein shows it to be a calcium-binding protein, Current Science, Vol.86, NO.6, 25 MARCH 2004

Abhishek Narain Singh, A105 Family Decoded: Discovery of Genome-Wide Fingerprints for Personalized Genomic Medicine, page 115-126, Proceedings of the International Congress on Personalized Medicine UPCP 2012 (February 2-5, 2012, Florence, Italy), Medimond Publisher, ScienceMED journal vol.3 issue 2, April 2012.

Gropp, William; Lusk, Ewing; Skjellum, Anthony (1996). "A High-Performance, Portable Implementation of the MPI Message Passing Interface". Parallel Computing. CiteSeerX 10.1.1.102.9485

Lin Xu et. al., Average Gene Length Is Highly Conserved in Prokaryotes and Eukaryotes and Diverges Only Between the Two Kingdoms, Mol Biol Evol (2006) 23 (6): 1107-1108. https://academic.oup.com/mbe/article/23/6/1107/1055387/Average-Gene-Length-Is-Highly-Conserved-in

Lina J. Karam, Ismail AlKamal, Alan Gatherer, Gene A. Frantz, David V. Anderson, Brian L. Evans (2009). "Trends in Multi-core DSP Platforms" . IEEE Signal Processing Magazine, Special Issue on Signal Processing on Platforms with Multiple Cores.

Majoros W, et al. (2004) TIGRscan and GlimmerHMM: two open-source ab initio eukaryotic gene finders, Bioinformatics 20, 2878-2879.

Salzberg, S. L.; Delcher, A. L.; Kasif, S.; White, O. (1998). "Microbial gene identification using interpolated Markov models". Nucleic Acids Research. 26 (2): 544–548. doi:10.1093/nar/26.2.544. PMC 147303 . PMID 9421513

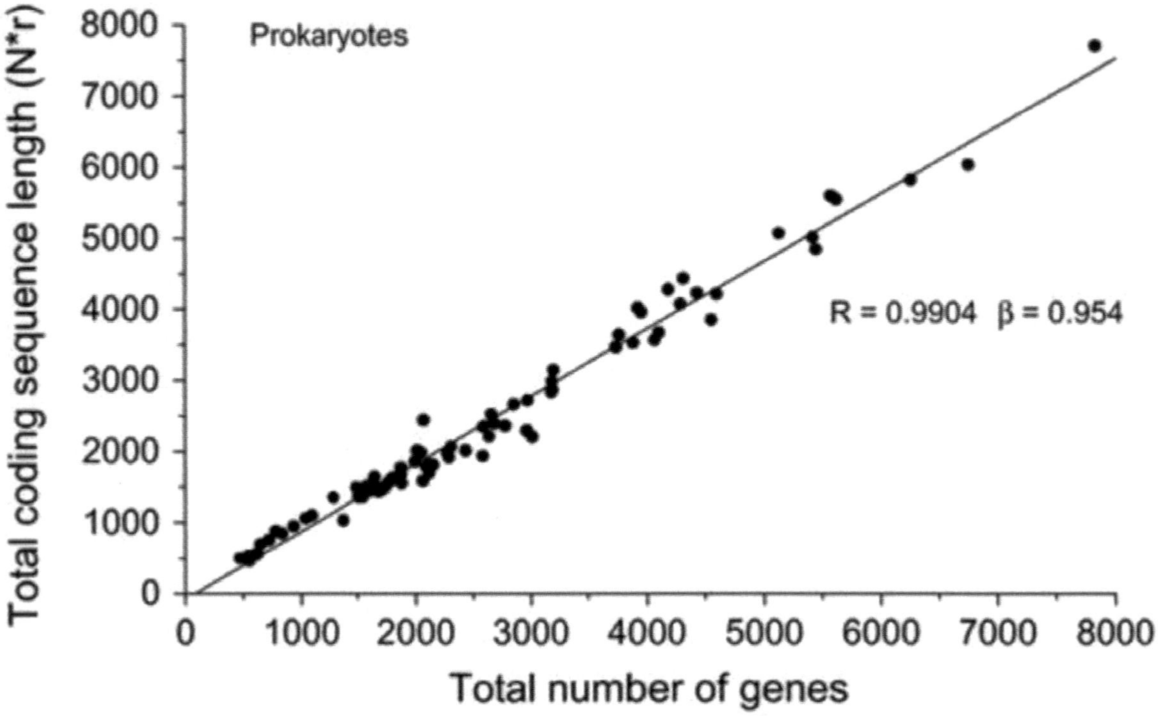

Figure 1. Xu et. al. 2006

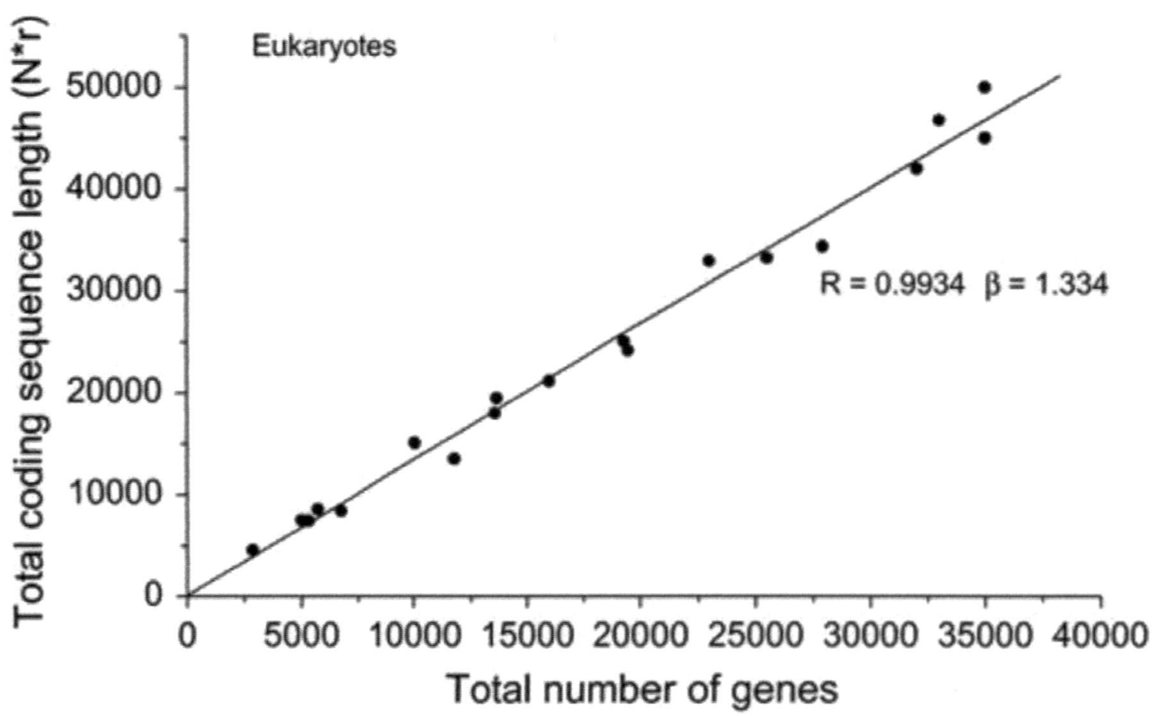

Figure 2. Xu et. al. 2006

SESSION

COMPUTATIONAL BIOLOGY, NOVEL ALGORITHMS, APPLICATIONS, AND TOOLS

Chair(s)

TBA

Analysis of Brain Scans from Live Zebrafish

R.S. Guidetti[1], C.D. Eichstaedt[2], J.L. Mustard[2], and N.W. Seidler[1,2]
[1]Department of Anesthesiology, UMKC -SOM/Saint Luke's Hospital, Kansas City, MO, USA
[2]Division of Basic Science, KCU-COM, Kansas City, MO, USA

Abstract - *Optical Coherence Tomography is a non-invasive technique useful in obtaining structural information on tissues and organs in animal models of human disease. The resolution of tissue scans are in the micrometer range, and this technique offers unique opportunities for biological experimentation, particularly as test subjects can be kept alive during and after imaging. OCT scanning is the optical analog of ultrasound imaging. This study demonstrates that this technique effectively assesses permanent physical changes to brain as a result of chronic ethanol ingestion. We describe the process of brain scanning, the ethanol treatment of test subjects, and the analysis of the data obtained. Adult zebrafish that were exposed to chronic levels of ethanol, followed by a significant washout period, exhibited changes in brain morphology consistent with hippocampal edema.*

Keywords: zebrafish, optical coherence tomography, brain scan, ethanol

1 Introduction

Damage to the CNS is a major complication of alcohol abuse. The health risks associated with excessive ethanol ingestion are many, including a greater prevalence (up to 5-fold increased risk) of post-operative complications, such as cognitive impairment [1]. Zebrafish is a useful model in studying alcohol-related disorders, and how alcohol impacts the effects of anesthesia. We previously proposed [2] that zebrafish is a potential animal model for post-operative cognitive dysfunction (POCD), a condition of serious concern for elderly patients [3]. The risk for POCD increases in individuals that excessively ingest ethanol [1], societally-accepted and potentially toxic non-nutrient agent. This study represents a preliminary examination of the effects of repeated ethanol exposure on adult zebrafish neurobehavior, using brain scans of live zebrafish. The brain scans were derived from optical coherence tomography (OCT), a non-invasive technology using interferometry of harmless light waves (λ=1,325nm). OCT can be used to scan zebrafish brain [4-6], which contains a forebrain structure with regions resembling those in humans (i.e. amygdala, hippocampus) [7]. This study examined the effects of chronic ethanol ingestion on telencephalon morphology, and describes the initial procedures that we used for analyzing the brain scans. We think that this technology can offer researchers a novel tool to ask, and answer, relevant questions regarding the function of an evolved brain.

1.1 Optical Coherence Tomography

OCT is a non-invasive imaging technique that uses light waves to take cross-section pictures of tissues. We used this technology to generate images of the brain in live zebrafish, namely the forebrain structure that contains the right and left hemispheres of the telencephalon and the olfactory bulbs. The axial resolution is 5.5 µm. OCT uses light from a matched pair of superluminescent diodes, which make up the reference and the sample beam, to obtain a reflectivity profile along the depth of the tissue. The use of near-infrared light of relatively long wavelength (i.e. λ = 1300nm) allows it to penetrate into the tissue. The backscattering of light waves from the tissue interferes with the reference beam, and the interference pattern is used to generate images to a depth of 3.5mm. The A-scan reflectivity profile of the light, derived from the reference and sample beams, scans one pixel width at a time down the Z-dimension incrementally along the X-dimension creating an image slice (i.e. B-scan). This process is repeated along the Y-dimension to generate a 3D image from stacked individual slices.

2 Methods and Results

Zebrafish (*Danio rerio*; wild-type AB) were bred and cared for using institutionally-approved standard protocols. Four female adult zebrafish were chosen from a cohort that was 537 days post fertilization (dpf) at the beginning of the study. Using the formula below, derived previously [2], where x is age of zebrafish in dpf and y is age equivalent in human years, we estimated the human age equivalent of the zebrafish used in this study to be 49 years old at the start of the study.

$$y = (0.425\,x^{0.757}) - 0.751 \qquad (1)$$

Two groups (chronic ethanol ingestion and control), each with two zebrafish, were exposed (or, mock-exposed) to ethanol as described below. We will refer to these four zebrafish as test fish in this report. The test fish were kept on a Z-Hab Mini system (pentairaes.com) under standard conditions. During treatment, they were kept in vessels containing freshly-prepared and filtered (0.45µm cellulose membranes) environmental water (EvH_2O), which consisted of 60mg/L of

Crystal Sea® Marinemix, which is a typical sea salt mixture, and 87.5μM sodium bicarbonate (pH ≈ 7.3).

2.1 Chronic Ethanol Ingestion

The test fish were exposed to ethanol (1% for 60min at 28°C), or mock-exposed (water bolus), for five consecutive days followed by a brief two-day washout period and then another five consecutive days of ethanol exposure. Using the age-equivalent formula (equation 1), this 12-day treatment period represents approximately two human years of alcohol ingestion. The upper limit in the United States for legal purposes is 0.1% blood alcohol concentration, indicating that the 1% ethanol exposure is well above the expected dosage that is shown to significantly impair both motor [8] and cognitive [9] function in humans. Exposure of zebrafish to 1% ethanol is similar to that used by other researchers [10, 11]. Blood vessels of the gills and skin efficiently absorb ethanol [12, 13] and within about 40min of immersion of the test fish in the treatment vessel the levels of ethanol in blood and brain will reach equilibrium with the concentration in the vessel [14, 15]. Gerlai and coworkers [13] demonstrated that 1% ethanol for 60min inhibits several behavioral endpoints, including aggression. In our study the 1% ethanol exposure for 60min was repeated ten times over a 12-day treatment period to represent chronic exposure to ethanol, translatable to chronic alcoholism in humans.

Our ethanol treatment protocol was conducted as follows. The test fish were given their morning feedings. After at least one hour, all four of the test fish were moved from the Zebrafish Facility to the OCT Lab. Four 1.0L-size spoutless beakers were prepared with pre-warmed (28°C) 720mL EvH_2O. The test fish were transferred by net to the beakers, covered with a small glass plate and placed in a lab incubator and kept for 20min at 28°C to acclimate to their surroundings. Then, an 80mL volume of EvH_2O or 10% ethanol (1% final concentration) was added to the respective beakers (two controls and two ethanol test fish). The glass-covered beakers were returned to the lab incubator and kept for 60min at 28°C. Black shields were used to visually isolate each fish. Following exposure, all fish were first transferred to a rinse beaker with 500mL pre-warmed (28°) EvH_2O prior to return to each group tank. The test fish were given their afternoon feeding at least one hour after returning to the Z-Hab Mini system. Due to the limited number of test fish in the 1.5L group tank, greenery was added to minimize aggressive behaviors.

2.2 Anesthetization Chamber

All four test fish were anesthetized, immobilized and scanned on the same day, one fish at a time. The test fish was transferred by net to the anesthetization chamber, which was a beaker (size: 150mL) with pre-warmed (28°C) EvH_2O (90mL), containing freshly prepared tricaine methane sulfonate (TMS) at a concentration of 100mg/L. The beaker was placed in a 200mL water bath (Peltier-controlled benchtop cooler) that was set to 21°C. The opaque cover was placed over the water bath and the beaker was allowed to reach 22°C (6min). The room temperature (rt) of the OCT Lab was 20°C. By visual observation, the test fish was immobile after approximately 1min. Upon reaching the designated temperature (22°C) in the anesthetization chamber, the test fish was quickly transferred by carefully decanting into the immobilization chamber.

2.3 Immobilization Chamber

The apparatus for positioning and immobilizing the test fish was setup as shown in Figure 1. EvH_2O (40mL) was added to the chamber and allowed to reach rt. After transferring to the immobilization chamber, the test fish was gently positioned manually into the spaces between the bristles (Figure 1), ensuring that the gill plates are unobstructed and submerged. The level of the water is adjusted to make sure that the entire head is just barely submerged for optimal OCT scanning.

Figure 1: Positioning Live Zebrafish
This photo of the immobilization chamber shows a glass culture dish (250mL-size) on a standard lab mat. In the center, there is a modified plastic base with smooth bristles that was custom cut from a commercial cosmetic product (i.e. hair detangler). A knotted rubber band kept the plastic base angled slight up from the bottom of the dish, to compensate for the downward curvature of the fish's head. Two large-sized microscope glass slides are placed to keep the plastic base submerged. The head of the test fish was positioned just past the bristles (the open lower area of the image).

2.3.1 Maintenance of Anesthesia

Maintaining optimal levels of anesthesia during OCT scanning ensures that the fish remains immobile and breathing calmly. EVH_2O and a concentrated solution of TMS were used by pipette to control water level and anesthesia.

2.3.2 Recovery from Anesthesia

Following scanning, the anesthetized test fish was transferred to a recovery beaker containing 120mL plain EvH_2O at rt. The beaker was place in a lab incubator (set to 28°C). The test fish was monitored for normal swim movement that should occur within two minutes of exposure to fresh EvH_2O. After a 40min acclimation period the test fish was then placed back into the group tank and returned to the Z-Hab Mini system.

2.4 Brain Scanning

The OCT device used in this study was purchased from ThorLabs (thorlabs.com), a Telesto series spectral domain OCT imaging system (TEL 1300V2-BU), equipped with a base unit, or engine, that consists of a superluminescent diode light source, scanning electronics, and a spectrometer for detection. The bandwidth is over 170nm with a 5.5μm axial resolution at an imaging depth of 3.5mm. The scanner is mounted on a focus block attached to an aluminum base with a fixed stage allowing for XY linear translation as well as rotational positioning capability. OCT scanning is the optical analog of ultrasound imaging.

The software allows for real-time 3D sectional imaging of the tissue simultaneously showing a standard, also real-time, video image of the subject, allowing for the experimenter to make adjustments to the stage in order to locate the target area. The image, as see in Figure 2, is brought into focus using coarse and fine knobs that control *z*-axis travel (40mm and 225μm per revolution, respectively) of the rigid scanner. Using this image, respiration rate can be determined by counting the number of gill openings per unit time.

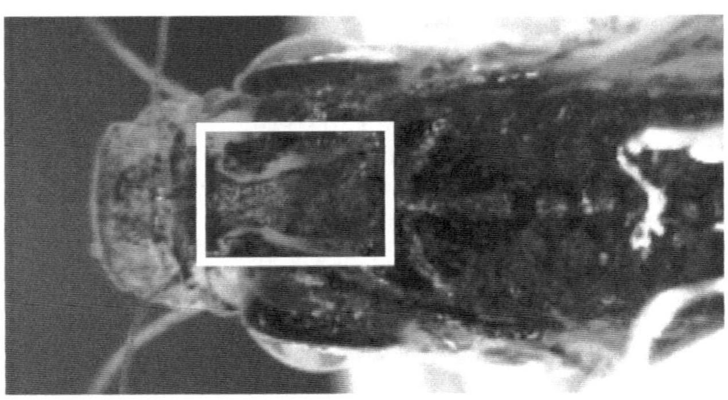

Figure 2: Locating the Target Telencephalon
This photo of a test fish shows the positioning of the head prior to the live brain scan with OCT. The *white box*, which defines the scan parameters, is drawn as a virtual marker above the area to be imaged. The zebrafish brain, which is elongated and segmented, exists with the forebrain (i.e. telencephalon/olfactory bulb), our target, situated between the eyes. The dimensions of the *white box* are kept constant, and a duplicate *white box* was utilized for each test animal. These photos were also used to determine head size.

The virtual rectangular box is drawn above the target area in the video image provided. As illustrated in Figure 3, the box is click-started at the lower left and click-ended at the upper right. The illustration in Figure 3 depicts the XYZ volume that is to be imaged. Operationally, once the box is drawn, its precise size can be adjusted and fixed. In our studies the X-, Y- and Z- dimensions were the same in all scans. The number of pixels along the XYZ dimensions was also set, and the same numbers were duplicated for each subject: 651, 769, and 640 pixels for X-, Y-, and Z-dimension, respectively. The intensity of the light source is adjusted to an optimal setting after choosing the target scanning speed: 28, 48, or 76 Hz. Next, the reference beam is adjusted to bring the 3D sectional image into proper focus. Once the 3D image is presented with the simultaneous video image of the subject's head, we determined the positional accuracy of the white box (Figure 2) and fine adjustments, using the translation stage, are performed to center the target telencephalon keeping both hemispheric lobes in view. Once all settings are defined and the image is optimized, the scans are acquired for permanent data collection. The scan acquisition time varies depending on the parameters set, particularly the pixel number and scan speed chosen. In our experiments the acquisition time varied between one and four minutes.

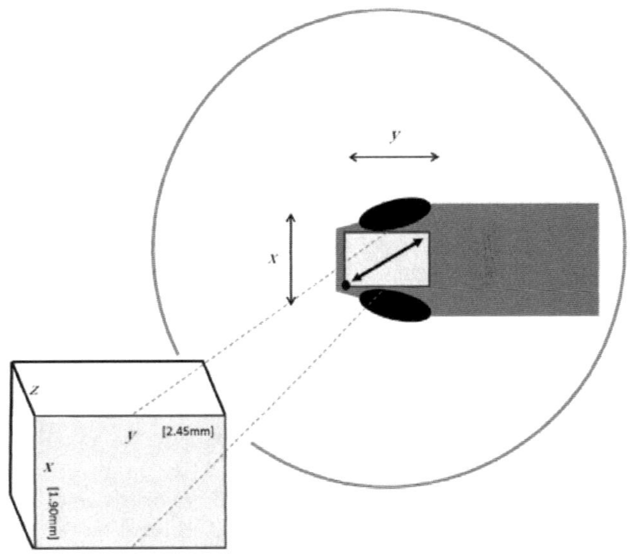

Figure 3: Target Zone for Scanning
The illustration schematically depicts a test fish (*dark gray polygon* with two *black beveled ovals*) representing the area of the body and eyes visible in the live video (see Figure 2) during preparation of the subject. The view here is looking down from the perspective of the optical device. The circle indicates the immobilization chamber that is initially manually adjusted, then fine-adjusted using the stage controls. The test fish is positioned left-to-right (the rostral-most point, or nose, is towards the left) in the immobilization chamber. The box is 1.90mm by 2.45mm, for X- and Y-dimensions respectively.

2.5 Resolution versus Image Stability

Resolution is a function of the chosen pixel density along one or more of the XYZ dimensions, such that the greater the number of pixels, the higher the resolution. Scan speed is another factor affecting image clarity, whereas the slower speeds (i.e. 28Hz) generate clearer images. Both of these parameters (high pixel density and low hertz setting) dramatically increase the length of time it takes to complete a single scan. Since we used live breathing zebrafish in our study, while the respiration rate was low in the anesthetized subjects, the repetitive gill movement caused a slight but disruptive shift in head (and therefore brain) position during the longer scans. This breathing-related head movement was less noticeable in the faster, less resolved, scan. This situation is a trade-off in balancing resolution over image stability, and remains an important issue going forward, as this technique is further utilized in experimentation.

2.6 Post-Scanning Analysis

We used the scans obtained at the highest speeds (i.e. 76Hz) in order to analyze the images with the least amount of instability due to head movement.

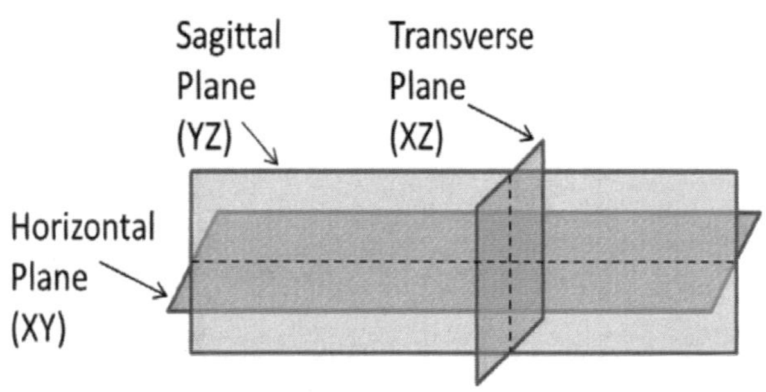

Figure 4: Anatomical Planes
The illustration defines the anatomical planes that represent the stacked slices of multiple 2D images acquired during scanning. Since we set the pixel density for the Y-dimension to be 769 (see discussion of Figure 3), we acquired 769 'transverse plane' slices. Likewise, the number of slices defined by the other two planes is identical to the corresponding number set as the pixel density. The transverse plane shows images proceeding in the rostral to caudal (nose to tail) direction. The horizontal plane shows images proceeding from the top to the bottom of the fish.

The length of the forebrain region (i.e. telencephalon and olfactory bulb) was calculated by examining sagittal plane images (Figure 4). To do this, we first determined the slice number (1 to 769) that defines the beginning of the olfactory bulb, the neuronal tissue just rostral to the telencephalon. And then, we determined the slice number corresponding to the junction between the end of the telencephalon and the beginning of the optic tectum. We then calculated the total number of slices that make up the forebrain, and then calculated the length in mm, using the relationship, 2.45mm per 769pixels in the Y-dimension.

We then determined the specific 'transverse plane' slice number that is one-quarter the total length of the forebrain, from the telencephalon-optic tectum junction. This target 'transverse plane' slice represented the center slice for further analysis (Figure 5). The group of stacked XZ slices was further processed, by measuring the width of the telencephalon lobes. The location chosen to measure the width was calculated by multiplying 25% of the forebrain length, and measuring down from the inner region of the skull. The measurements were performed by two different experimenters and then averaged. Additionally, the widths of the two control fish (and two ethanol-exposed) were averaged, normalized and compared.

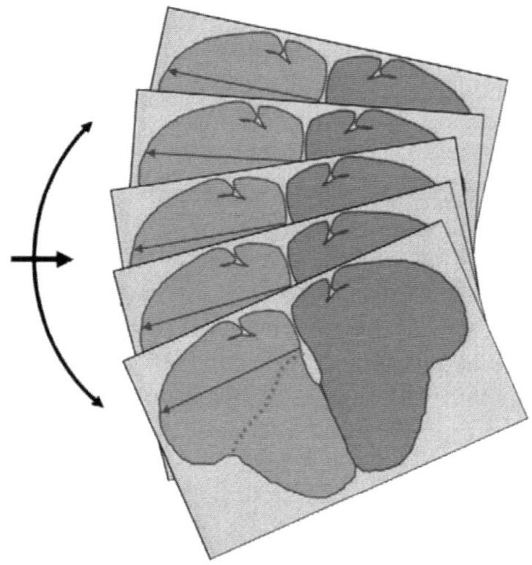

Figure 5: Designated 'transverse plane' XZ Slices
This illustration shows the center target XZ slice (*left horizontal arrow*) as well as the additional contiguous slices (minimum of 12 in each direction, shown by *curved arrow*) in the rostral and caudal directions. The bi-lobed telencephalon is schematically shown in *medium and dark gray*. The *dotted line* demarcates the upper pallial and lower sub-pallial regions. The upper pallial area consists of regions similar to that found in humans, namely the amygdala, isocortex and hippocampus [16]. The *small arrows* shown in the left lobes in the stacked slices indicate distances that were measured and compared.

2.7 Normalization of Data

Post-scanning analysis involved normalizing the lobular telencephalon widths for each subject. To do this, we measured the head size using the points immediately rostral to the optic socket from the video image provided during scanning (see Figure 2). The two control head sizes were averaged and then all data was normalized to that value. The head sizes were accurately assessed since each recorded video image contained the 'white box' described in Figure 2, which was 1.9mm in the orientation of the head size measurement.

2.8 Comparison of Lobular Widths

The zebrafish telencephalon is made up of two lobes as shown in the schematic slices in Figure 5. In each of these slices, the *small arrows* represent the measured distance defined as lobular widths. The normalized data of the lobular widths of the pallial telencephalon are shown in Figure 6. Control zebrafish are compared with those chronically exposed to ethanol. The data were compared using a paired *t*-test, which demonstrated a significant difference between the two groups ($p < 0.001$; two-tailed).

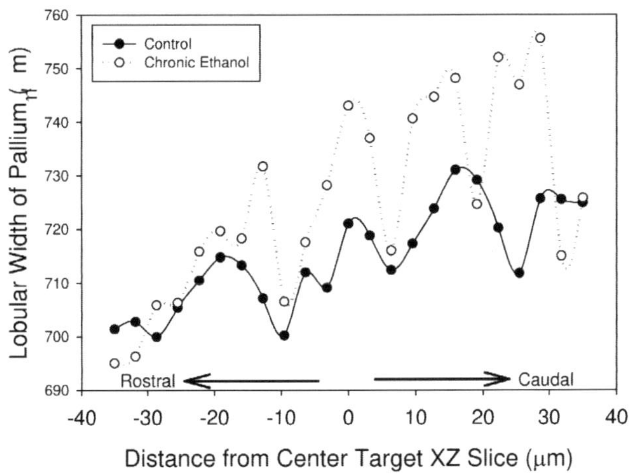

Figure 6: Telencephalon Lobes from Control Zebrafish and those Chronically Ethanol-Exposed The graph compares the lobular widths measured from the brain scan XZ slices of control zebrafish (*solid line* and *filled circles*) with those measured from brain scan XZ slices of zebrafish chronically exposed to ethanol (*dotted line* and *open circles*). Each symbol represents the mean from two zebrafish (after first averaging the measurements obtained from two independent experimenters). The data points were linked using spline lines to better visualize the surface contour of the telencephalon. The center target XZ slice (*vertical dashed line*) is given the value zero, and the distance of each slice from the center target is given in µm, progressing in the rostral and caudal directions.

The data was also compared using an *f*-test to compare variances between the two groups. The results also indicated a significant difference ($p < 0.01$). These observations suggest that the telencephalon in the zebrafish that were chronically exposed to ethanol was edematous and exhibited an abnormal uneven surface contour.

3 Conclusions

Since this ethanol-induced physical change to the region of the telencephalon that contains hippocampal and amydala nuclei, we would expect that the zebrafish chronically exposed to ethanol would behavioral challenges associated with spatial navigation and emotional responses to novel environment or presence of a predator. Permanent behavioral issues, such as aberrations in navigating through a familiar or unfamiliar tank or environment, would be expected. We intend to examine these neurophenotypic behaviors. Our findings indicate that the telencephalon in the zebrafish that were chronically exposed to ethanol was edematous, consistent with the long-standing hypothesis of overhydration [17] that occurs during the withdrawal period of alcoholism. Our subjects were tested 11 days after alcohol withdrawal. Curiously, the hippocampus appears particularly susceptible to the damaging effects of chronic ethanol ingestion, as evidenced by a significant volume loss in active alcoholic human subjects relative to controls [18]. The area of the telencephalon that we measured contains the hippocampal region. We think that there was an alcohol-induced pathology to this area that proceeded to experience an overhydration during the withdrawal period, resulting in an edematous-structure that exhibited an abnormal uneven surface contour.

4 References

[1] Hudetz JA, Iqbal Z, Gandhi SD, Patterson KM, Hyde TF, Reddy DM, Hudetz AG, Warltier DC. Postoperative cognitive dysfunction in older patients with a history of alcohol abuse. Anesthesiology. 2007;106 (3):423-30.

[2] McElroy B, Mustard J, Kamran S, Jung C, Bakken K and Seidler NW (2016) Modeling post-operative cognitive dysfunction in zebrafish. Advances in Alzheimer's Disease, 5, 126-141.

[3] Rundshagen I. Postoperative cognitive dysfunction. Dtsch Arztebl Int. 2014 ;111 (8):119-25.

[4] Zhang Z, Zhu B, Ge W, Genetic analysis of zebrafish gonadotropin (FSH and LH) functions by TALEN-mediated gene disruption. Mol. Endocrinol. 2015; 29 (1):76-98.

[5] Zhang J, Ge W, Yuan Z, In vivo three-dimensional characterization of the adult zebrafish brain using a 1325 nm spectral-domain optical coherence tomography system with the 27 frame/s video rate. Biomed. Opt. Express 2015;6 (10): 3932-3940.

[6] Rao KD, Alex A, Verma Y, Thampi S, Gupta PK, Real-time in vivo imaging of adult zebrafish brain using optical coherence tomography. J. Biophoton. 2009;2 (5):288-291.

[7] Mueller T, Dong Z, Berberoglu MA, Guo S. The Dorsal Pallium in Zebrafish, Danio rerio (Cyprinidae, Teleostei). Brain research. 2011;1381:95-105.

[8] Nuotto EJ, Korttila KT (1991) Evaluation of a new computerized psychomotor test battery: effects of alcohol. Pharmacology & Toxicology 68: 360–365.

[9] Dry MJ, Burns NR, Nettelbeck T, Farquharson AL, White JM. Dose-related effects of alcohol on cognitive functioning. PLoS One. 2012;7(11):e50977

[10] Chacon DM, Luchiari AC A dose for the wiser is enough: the alcohol benefits for associative learning in zebrafish. Prog Neuropsychopharmacol Biol Psychiatry. 2014;53:109-15.

[11] Li X, Li X, Li YX, Zhang Y, Chen D, Sun MZ, Zhao X, Chen DY, Feng XZ. The Difference between Anxiolytic and Anxiogenic Effects Induced by Acute and Chronic Alcohol Exposure and Changes in Associative Learning and Memory Based on Color Preference and the Cause of Parkinson-Like Behaviors in Zebrafish. PLoS One. 2015;10(11):e0141134.

[12] Lockwood B, Bjerke S, Kobayashi K, Guo S. Acute effects of alcohol on larval zebrafish: a genetic system for large-scale screening. Pharmacology, Biochemistry, Behaviour. 2004; 77:647–654.

[13] Gerlai R, Lahav M, Guo S, Rosenthall A. Drinks like a fish: zebra fish (Danio rerio) as a behavior genetic model to study alcohol effects. Pharmacology Biochemistry and Behavior. 2000; 67:773– 782)

[14] Dlugos C, Rabin R. Ethanol effects on three strains of zebrafish: model system for genetic investigations. Pharmacology, Biochemistry, and Behavior. 2003; 74(2):471–480.

[15] Ryback R, Percarpio B, Vitale J. Equilibration and metabolism of ethanol in the goldfish. Nature. 1969; 222:1068–1070.

[16] Mueller T, Dong Z, Berberoglu MA, Guo S. The dorsal pallium in zebrafish, danio rerio (Cyprinidae, Teleostei). Brain research. 2011;1381:95-105.

[17] Lambie DG. Alcoholic brain damage and neurological symptoms of alcohol withdrawal--manifestations of overhydration. Med Hypotheses. 1985;16 (4):377-88.

[18] Agartz I1, Momenan R, Rawlings RR, Kerich MJ, Hommer DW. Hippocampal volume in patients with alcohol dependence. Arch Gen Psychiatry. 1999;56 (4):356-63.

Three-State Protein Stability Prediction from Sequence-Based Features

Jose A. Guevara-Coto[1], Charles E. Schwartz[2], and Liangjiang Wang[1]
[1]Department of Genetics and Biochemistry, Clemson University, Clemson, SC 29634, USA
[2]J.C. Self Research Institute of Human Genetics, Greenwood Genetic Center, Greenwood, SC 29646, USA

Abstract - *Amino acid substitutions can have significant and deleterious effects on proteins. Prediction of the effects of substitutions on protein stability has been explored, but many studies make use of structure-based features, which are not available for all proteins. In this study, we have developed a sequence-based SVM model for three-state protein stability prediction. This model used features extracted from the primary sequence, and feature selection identified the most informative feature set for model construction. We evaluated this model with an independent test dataset, and obtained the accuracy of 70.52% with 61.20% sensitivity and 79.84% specificity. Our results suggest that sequence features contain sufficient information for accurate prediction of three-state protein stability changes caused by amino acid substitutions.*

Keywords: Amino acid substitutions, three-state protein stability prediction, sequence features, support vector machines

1 Introduction

Disease-causing sequence changes have been identified for many human genes [1]. The type of changes varies in nature, affecting multiple mechanisms from RNA processing to post-translational modifications. However, it has been identified that the most common effect of these changes is on protein stability [1–3]. An analysis of human single nucleotide polymorphisms (SNPs) revealed that ~80% of disease-causing amino acid substitutions affect protein stability [4]. In humans, ~60% of single nucleotide missense mutations in coding regions account for monogenic diseases based on the Human Gene Mutation Database [1]. Thus, understanding the effects of sequence variations on proteins is an important task.

Variations in the amino acid sequence can impact the physicochemical characteristics of a polypeptide. These sequence changes may alter the biochemical properties of the protein, resulting in modified structural characteristics with deleterious effects. The notion of predicting the effect of amino acid substitutions on protein stability has been previously explored [5, 6]. However, our approach, similar to [7, 8] and more recently [9], proposes to focus on sequence-based features and avoid using any features based on the protein structure.

The development of a three-state protein stability predictor was previously reported [10]. However, the approach made use of structural features which require a known protein structure. This can be a limiting factor as not all proteins have structural information available. Although computational methods are available for protein structure prediction [11], they are not as accurate in resolution as the experimental determination of a protein structure. The limitation of previous works in the use of structural features may be circumvented by encoding the instances with features derived completely from amino acid sequence.

In this study, we have developed a new three-state protein stability predictor based on sequence features. The importance of the sequence features in model performance was examined using two feature selection methods capable of reducing and ranking variables, random forests and recursive feature elimination. Our three-state protein stability predictor based on Support Vector Machines (SVMs) showed performance comparable to the currently available methods using structural information. The results suggest that sequence features can provide useful information for predicting the effect of amino acid substitutions on protein stability.

2 Methods

2.1 Data acquisition

The protein dataset was derived from that used by Capriotti et al [10] and later modified by Folkman et al [12]. We modified this dataset to contain 68 unique protein entries whose sequences were obtained in FASTA format from the Protein Data Bank (PDB)

[13]. The sequences were subsequently subjected to redundancy reduction with a threshold of 85% sequence similarity using BlastClust. This assured the uniqueness of each sequence and eliminated the presence of multiple chains for a single entry. This process also resulted in the condensation of small sequences into larger clusters. The resulting dataset contained 1,332 non-redundant instances (henceforth s1332) with information about the PDB identifier, the wild-type position, the mutated amino acid, the substituted position, and the free energy change ($\Delta\Delta G$), which was used to determine the class label of the instance. In this study, we had three classes or states: decreased stability (DS), increased stability (IS), and no significant change (NC). The $\Delta\Delta G$ thresholds used in this study were as follows: DS if the effect of an amino acid substitution on the protein stability change $\Delta\Delta G \leq -0.5$ kcal/mol, IS if $\Delta\Delta G \geq 0.5$ kcal/mol, or NC if -0.5 kcal/mol $\leq \Delta\Delta G \leq 0.5$ kcal/mol. For building an independent test dataset, we selected the s238 sequence set, and subjected it to the same process as the s1332 dataset. This test dataset contained unique entries that were not included in the training dataset s1332, and consisted of variants representing the three different states.

2.2 Sequence-based features

The instances in the s1332 dataset were encoded with a total of 31 sequence features using the R package "Peptides" [14]. This study used various physicochemical and biochemical features as defined by ExPASy in ProtParam (http://web.expasy.org/protparam/protparam-doc.html). These features include molecular mass, amino acid composition, charge, and aliphatic index, defined as the relative volume occupied by aliphatic side chains (Alanine, Valine, Isoleucine, and Leucine). Other features include: the Kidera factors, which comprise 10 features derived from 188 physical properties; three different hydrophobicity indices, obtained with three different scales (Fasman, Bull, Chothia); instability index, based on dipeptide composition; and the Boman index, which indicates the potential of a peptide to bind to the membrane or other proteins.

2.3 Model construction

In this study, the multiclass protein stability model was constructed using Support Vector Machines (SVMs) with sequence-based features. SVMs were shown to produce well-performing models for protein stability prediction in previous studies [7,8]. In its essence, an SVM model defines a separation hyperplane that divides the space into two distinct halves. Based on the sign given by the $f(x)$, a point will be assigned to a given side of the hyperplane. If $f(x)>0$, a point will be assigned to the positive side of the hyperplane. The soft margin can increase the performance of the classifier when compared to hard margins, and this is achieved by allowing misclassification of some points [15]. The use of soft margins in SVMs comes as a response to the fact that not all data is linearly separable, which is especially true with biological data [15]. The third component refers to kernels, which can be used to make the calculation process more efficient, especially in feature spaces of high dimensionality. The radial basis function (RBF) kernel was used to construct the SVM model in this study. RBF is one of the most widely used kernel functions in model development and often performs well on biological data [15].

The *caret* and *e1071* packages [16,17] available in R were used to construct the SVM model. We examined multiple options for model construction, including multiclass SVM models, kernel functions, and performance metrics. The Receiver Operating Characteristic (ROC) curve, specifically the area under the curve (ROC-AUC), was calculated using the package pROC [18]. The SVM model was also compared with two different Random Forest (RF) classifiers, a single RF and an ensemble of RFs. The RF learning algorithm was previously shown to perform well for protein stability prediction [19].

2.4 Feature selection

Feature or variable selection for a classification problem can have two main objectives: (1) to identify highly important variables that are related to the response variable, and (2) to reduce the feature space to improve the prediction of the class label [20]. An efficient feature selection method can not only achieve these objectives, but may also combine important components such as determining importance thresholds, variable ranking and stepwise introduction of variables into the feature set [21, 22]. In this study, we implemented a Random Forest (RF) based feature ranking method [21]. RFs are built upon decision trees, with every node being a condition on a variable. We also tested the recursive feature elimination method [22], in which the lowest 20% features were eliminated in each round to determine the most important sequence-based features.

2.5 Model performance evaluation

Model performance was evaluated using the R package performanceEstimation [23]. We used the tenfold cross-validation method. Briefly, this method consists of partitioning the data into different folds where one-fold is used for testing and the remaining folds are used for training. We also used the independent test dataset s238 to evaluate model performance. The following metrics were used in this study to measure model performance:

$$\text{Accuracy} = \frac{TP+TN}{TP+TN+FP+FN} \quad (1)$$

$$\text{Sensitivity} = \frac{TP}{TP+FN} \quad (2)$$

$$\text{Specificity} = \frac{TN}{TN+FP} \quad (3)$$

$$\text{MCC} = \frac{TP \times TN - FP \times FN}{\sqrt{(TP+FP)(TP+FN)(TN+FP)(TN+FN)}} \quad (4)$$

The accuracy provides the information regarding the true positives (TP) and true negatives (TN) in the total of the dataset, which also includes the identified false positives (FP) and false negatives (FN). Sensitivity or true positive rate refers to the proportion of positive instances that were properly identified in each class, whereas specificity or true negative rate is the number of negative instances identified as such. The Matthews Correlation Coefficient (MCC) was also used in this study to measure the performance of multi-class models. For a model with three classes (A, B and C), the values for TP, TN, FP and FN can be calculated as described previously [24]:

$$TP = TP_A + TP_B + TP_C \quad (5)$$

$$TN = TN_A + TN_B + TN_C \quad (6)$$

$$FP = FP_A + FP_B + FP_C \quad (7)$$

$$FN = FN_A + FN_B + FN_C \quad (8)$$

3 Results

3.1 Prediction of three-state protein stability changes

To develop an accurate model for three-state protein stability prediction, we tested two widely used machine learning algorithms, Support Vector Machine (SVM) and Random Forest (RF). The SVM and RF models were constructed using 31 sequence features. As shown in **Table 1**, the SVM method outperformed the RF learning algorithm for predicting protein stability changes. The SVM model that was fine-tuned with the training parameters (C = 15, *gamma* = 0.40) achieved higher performance measures than the RF models in the tenfold cross-validation. In this study, we constructed two RF models: a single RF model and an ensemble comprising of three RFs (RF-E). It was previously shown that an ensemble could result in improved model performance [25]. However, the RF ensemble achieved similar performance measures as the single RF model for protein stability prediction (**Table 1**). We thus selected the SVM model for further analyses.

3.2 Selection of relevant sequence features for model construction

Feature selection was conducted to determine the impact of sequence features in the model's ability to discriminate the three protein stability states, and to potentially enhance the model performance. The first feature selection method ranked the importance of the variables (sequence features) using the Gini index [22,26]. The approach based on *Boruta* [26] sets the variable selection threshold based on the value of shadow attributes, which are shuffled copies of all attributes to create randomness, culminated in the reduction of the feature set from 31 to 12 features. The second method, recursive feature elimination, identified a set of 7 features. Interestingly, the two feature selection methods identified several common features, including the hydrophobicity indices of Fasman and Chothia, some of the Kidera factors, and the aliphatic index. The first method also identified additional features associated with the putative overall effect of amino acid substitutions on the peptide, such as the isoelectric point.

However, the SVM models constructed with the selected features did not show improved performance in the tenfold cross-validation (**Table 2**). The SVM model using all the 31 sequence features (SVM_Full) achieved higher accuracy, ROC-AUC and MCC than the two models after feature selection (SVM_12 and SVM_7). One possibility was that the model SVM_Full might be slightly overfitted. To examine this possibility, we further compared the model performance using an independent test dataset.

3.3 Model performance evaluation using an independent test dataset

When compared using an independent test dataset (s238), the SVM models constructed with the selected features appeared to give slightly better performance measures than the model with the full feature set (**Table 3**). In particular, the SVM model using the 12 selected features (SVM_12) achieved slightly better ROC-AUC and MCC than the other two models (SVM_Full and SVM_7). The results suggest that the 12-feature set might be optimal for predicting three-state protein stability changes. Using the full set of 31 sequence features could cause model overfitting, whereas the 7-feature set might not provide sufficient information for prediction.

4 Conclusions

In this study, we have developed a new model for three-state prediction of protein stability changes caused by amino acid substitutions. The SVM model was built with sequence-based features. Feature selection identified 12 features for accurate protein stability prediction. We further validated the predictive performance of the model using an independent test dataset. Our results suggest that sequence features can provide sufficient information for predicting the effect of amino acid substitutions on protein stability.

Table 1. Performance of the SVM and RF models based on tenfold cross-validation. The models were constructed using 31 sequence features.

Model	AUC	MCC	Accuracy	Sensitivity	Specificity
SVM	**0.8972**	**0.7038**	**0.8659**	**0.8411**	**0.8907**
RF	0.7654	0.5360	0.7938	0.6907	0.8453
RF-E	0.7666	0.5163	0.7826	0.6814	0.8348

Table 2. Performance of the SVM models after feature selection based on tenfold cross-validation. SVM_Full was constructed using all the 31 sequence features; SVM_12 was constructed with 12 selected features; and SVM_7 was constructed with 7 selected features.

Model	AUC	MCC	Accuracy	Sensitivity	Specificity
SVM_Full	**0.8972**	**0.7038**	**0.8659**	**0.8411**	**0.8907**
SVM_12	0.8827	0.6115	0.8204	0.7830	0.8578
SVM_7	0.8439	0.5529	0.7902	0.7445	0.8360

Table 3. Predictive performance of the SVM models on an independent test dataset. SVM_Full was constructed using all the 31 sequence features; SVM_12 was constructed with 12 selected features; and SVM_7 used 7 selected features.

Model	AUC	MCC	Accuracy	Sensitivity	Specificity
SVM_Full	0.7423	0.3682	0.6647	0.5731	0.7563
SVM_12	**0.7477**	**0.4555**	0.7052	0.6120	**0.7984**
SVM_7	0.7367	0.4452	**0.7092**	**0.6235**	0.7950

5 References

[1] T. Alber, "Mutational effects on protein stability," *Annual Review Biochemistry*, no. 58, pp. 765–798, 1989.

[2] V. Ramensky, P. Bork, and S. Sunyaev, "Human non-synonymous SNPs : server and survey," *Nucleic Acids Research*, vol. 30, no. 17, pp. 3894–3900, 2002.

[3] Z. Wang and J. Moult, "SNPs , Protein Structure , and Disease," *Human Mutation*, vol. 270, pp. 263–270, 2001.

[4] P. D. Stenson *et al.*, "The Human Gene Mutation Database: 2008 update," *Genome Medicine.*, vol. 1, no. 1, pp. 1–6, 2009.

[5] L. Huang, M. M. Gromiha, and S. Ho, "Structural bioinformatics iPTREE-STAB : interpretable decision tree based method for predicting protein stability changes upon mutations," *Bioinformatics*, vol. 23, no. 10, pp. 1292–1293, 2007.

[6] E. Capriotti, P. Fariselli, R. Calabrese, and R. Casadio, "Protein Structure and Function Predicting protein stability changes from sequences using support vector machines," *Bioinformatics*, vol. 21, pp. 54–58, 2005.

[7] J. Cheng, A. Randall, and P. Baldi, "Prediction of Protein Stability Changes for Single-Site Mutations Using Support Vector Machines," *Proteins: Structure, Function, and Bioinformatics*, vol. 62, no. 4, pp. 1125–1132, 2006.

[8] S. Teng, A. K. Srivastava, and L. Wang, "Sequence feature-based prediction of protein stability changes upon amino acid substitutions," *BMC Genomics*, vol. 11, no. Suppl 2, p. S5, 2010.

[9] L. Folkman, B. Stantic, and A. Sattar, "Sequence-only evolutionary and predicted structural features for the prediction of stability changes in protein mutants," *BMC Bioinformatics*, vol. 14, no. 2, p. S6, 2013.

[10] E. Capriotti, P. Fariselli, and R. Casadio, "I-Mutant2 . 0 : predicting stability changes upon mutation from the protein sequence or structure," *Nucleic Acids Research*, vol. 33, no. suppl 2, pp. W306–W310, 2005.

[11] S. Raman *et al.*, "Prediction Report Structure prediction for CASP8 with all-atom refinement using Rosetta," *Proteins: Structure, Function, and Bioinformatics*, vol. 77, no. S9, p. 89–99., 2009.

[12] L. Folkman, B. Stantic, and A. Sattar, "Feature-based multiple models improve classification of mutation-induced stability changes," *BMC Genomics*, vol. 15, no. 4, p. S6, 2014.

[13] H. M. Berman *et al.*, "The Protein Data Bank," *Nucleic Acids Research*, vol. 28, no. 1, pp. 235–242, 2000.

[14] D. Osorio, P. Rondon-Villarreal, and R. Torres, "Peptides: A Package for Data Mining of Antimicrobial Peptides," *The R Journal*, vol. 7, no. 1, pp. 4–14, 2015.

[15] A. Ben-Hur, C. S. Ong, S. Sonnenburg, B. Schölkopf, and G. Rätsch, "Support vector machines and kernels for computational biology," *PLoS Computational Biology*, vol. 4, no. 10, 2008.

[16] M. Kuhn *et al.*, "Caret package," *Journal of Statistical Software*, vol. 28, no. 5, pp. 1–26, 2008.

[17] E. Dimitriadou, K. Hornik, F. Leisch, D. Meyer, A. Weingessel, and M. F. Leisch, "Package 'e1071,'" *R Software Package, avaliable at http://cran.rproject.org/web/packages/e1071/index. html*. pp. 1–62, 2009.

[18] X. Robin *et al.*, "pROC: an open-source package for R and S+ to analyze and compare ROC curves," *BMC Bioinformatics*, vol. 12, p. 77, 2011.

[19] Y. Li and J. Fang, "PROTS-RF : A Robust Model for Predicting Mutation- Induced Protein Stability Changes," *PLoS One*, vol. 7, no. 10, p. e47247, 2012.

[20] L. Zhang and P. Nagaratnam, "Random Forests with ensemble of feature spaces," *Pattern Recognition*, vol. 47, no. 10, pp. 3429–3437, 2014.

[21] R. Genuer, J. Poggi, and C. Tuleau-Malot, "Variable selection using Random Forests," *Pattern Recognition Letters*, vol. 31, no. 14, pp. 2225–2236, 2010.

[22] M. B. Kursa and W. R. Rudnicki, "Feature Selection with the Boruta Package," *Journal of Statistical Software*, vol. 36, no. 11, pp. 1–13, 2010.

[23] R. Díaz-Uriarte and S. A. De Andrés, "Gene selection and classification of microarray data using random forest," *BMC Bioinformatics*, vol. 7, no. 1, p. 3, 2006.

[24] L. Torgo, "An infra-structure for performance estimation and experimental comparison of predictive models in R," *arXiv Prepr. arXiv1412.0436*, 2015.

[25] F. Sebastiani, "Machine Learning in Automated Text Categorization," *ACM Computational Surveys*, vol. 34, no. 1, pp. 1–47, 2002.

[26] M. B. Kursa and W. R. Rudnicki, "Wrapper Algorithm for All Relevant Feature Selection," *R Software Package, available at https://m2.icm.edu.pl/boruta/*. 2016.

Using Computerized ECG Measurements in the Bayesian Analysis of Heart Disease

Robert A. Warner, MD*

Tigard Research Institute
12228 SW Chandler Drive
Tigard, OR, 97224 USA
hillwarner@frontier.com

Keywords: systolic dysfunction, ECG data, prior probability

Regular Research Paper

Abstract

The study utilized 1096 sets of diagnostic data from patients who reported to emergency departments with acute shortness of breath. Left ventricular dysfunction (LVSD) was defined as an echocardiographic left ventricular ejection fraction <50% and the diagnostic sensitivities for LVSD at 98% specificity for two tests for LVSD – brain natriuretic peptide (BNP) and the electronically recorded third heart sound (S3) were determined. For both the BNP and the S3 tests, an automated ECG criterion - duration of the QRS complex >120 ms. - identified subgroups with higher prior probabilities of LVSD than the total population. As expected from Bayesian principles, both the BNP and S3 tests exhibited better performances for detecting LVSD in the higher prior probability subgroups than in the total population. Computerized measurement of ECG QRS duration can assess the prior probability of LVSD and help improve the performances of specific tests for LVSD.

1.0 Introduction

The electrocardiogram (ECG) is an inexpensive, convenient and widely available test that is often used in patients with known or suspected heart disease. The nature of the diagnostic information provided by the ECG is very useful for detecting arrhythmias and ischemic heart disease and for suggesting the presence of cardiac chamber enlargement. In contrast, ECG data are considered to be unsuitable for directly detecting hemodynamic abnormalities such as left ventricular systolic dysfunction (LVSD), a condition that is often associated with disabling and potentially lethal heart failure.

However, in keeping with the principles of Bayesian statistics, I conjectured that ECG data could be used to augment tests that are specifically intended to detect LVSD. This is because heart disease often has multiple manifestations whose presence can be suggested by different types of tests, including the ECG. Conversely, most patients who are free of heart disease have normal findings on multiple cardiological tests. Therefore, the prevalence of LVSD is likely to be higher in patients who have even nonspecific ECG evidence for heart disease than it is in patients who lack such evidence. In the terminology of Bayesian analysis, the presence of ECG abnormalities

in a patient increases the prior probability that LVSD is also present in that patient.
In this study, I selected the duration of the QRS complex as the ECG parameter to determine the prior probability that LVSD is present in each member of a population of patients. The QRS complex is the portion of the ECG signal inscribed during electrical activation, i.e. depolarization, of the ventricles. The QRS duration in ms. is routinely reported in ECG interpretations and is a highly reproducible measurement whether it is made visually or algorithmically. The upper limit of normal of the QRS duration is 100 ms., borderline values are 100 to 120 ms. and abnormal values exceed 120 ms. The use of QRS duration is a particularly relevant parameter in the present context because a commonly used type of treatment for LVSD with heart failure – cardiac resynchronization therapy - has been found to be particularly effective if the patient's QRS duration is abnormally prolonged.[1]

A test that is often used to identify LVSD directly is the echocardiogram. The most useful echocardiographic parameter for this purpose is the left ventricular ejection fraction. The left ventricular ejection fraction expresses the percentage of the left ventricular end-diastolic blood volume that is ejected from the ventricle during a single left ventricular systolic contraction. The mean normal value of this parameter is about 65% and any value of the ejection fraction less than 50% is considered to indicate that LVSD is present.[2]

Although the echocardiogram is a definitive test for LVSD, it is expensive, not always readily available for the evaluation of patients and requires special expertise to record and interpret. Therefore, other less expensive and more readily available tests are routinely used to detect LVSD and the heart failure that is often associated with it. One of these is a blood test that measures brain natriuretic peptide (BNP). BNP is elevated in LVSD with heart failure and, in the appropriate clinical context, values of BNP >500 pg/ml are considered to be strong evidence of LVSD with heart failure.[3] Also often associated with LVSD is the presence of a third heart sound (S3).[4] The S3 is a low pitched sound that originates in the ventricle and occurs in early diastole. The S3 can be heard with a stethoscope or can be detected and recorded using an electronic sound sensor applied to the left side of the chest.[5] An advantage of recording the S3 electronically is that its strength can be quantified using the amplitude and frequency that the sound exhibits in a given patient.

Based on the above considerations, I tested the following hypotheses:

- The prevalence, i.e. the prior probability, of LVSD is greater in subgroups of patients with prolonged ECG QRS complexes than it is in the general population.
- In accordance with the principles of Bayesian statistics, the diagnostic performances of the BNP and S3 tests for detecting LVSD are better in populations of patients with a higher prevalence of LVSD than in populations with a lower prevalence of LVSD.

2.0 Materials and Methods

2.1 Selection of Patients

I studied a total of 1096 sets of data from a convenience sample of patients (mean age 61 years, 32% women) who had presented with acute shortness of breath to the emergency department of one of several metropolitan hospitals.

2.2 Diagnostic Tests

In each case, the left ventricular ejection fraction was measured by echocardiography within 24 hours of each patient's arrival at the hospital. At the time of arrival at the hospital, 432 patients also had an ECG and electronically recorded heart sounds (Audicor™, Inovise Medical, Inc. Portland, Oregon, USA) and in 374 patients, BNP was measured. The unit of acoustical strength of the S3 was the "display value" - a proprietary parameter that utilizes both the amplitude and the frequency of the recorded sound. Blood levels of BNP were expressed in pg./ml.

2.3 Analysis of the Data

LVSD was considered present if the patient's left ventricular ejection fraction was <50%. Receiver-operating characteristic curves were used to determine the diagnostic performances of the S3 and BNP in the entire group of patients as well as in the subgroup with ECG QRS duration >120 ms. For both the S3 and the BNP data, chi square analysis was used to compare diagnostic sensitivities at 98% specificity in the entire group of patients vs. the patients with QRS duration >120 ms.

3.0 Results

3.1 BNP Blood Test

Table 1 illustrates the importance of considering the duration of the ECG QRS complex when using BNP to detect LVSD. Table 1 shows that in the entire group of patients in which BNP was measured, the prevalence of LVSD was 48.9% and in the subgroup with prolonged QRS duration, the prevalence of LVSD was 82.8%. In the entire, lower prevalence group of patients, the diagnostic sensitivity at 98% specificity of BNP for detecting LVSD was 10.9%. In the subgroup with prolonged QRS duration, the diagnostic sensitivity of BNP for detecting LVSD rose to 46.8% and this difference was highly statistically significant. Table 1 also shows that in the entire group of 374 patients compared to the subgroup with prolonged QRS duration, the threshold values of BNP needed to attain 98% diagnostic specificity were 1740 and 407 pg./ml., respectively. Furthermore, the ratio of true positive to false positive BNP test results was 5.2 in the entire group and increased to 112.5 in the subgroup with prolonged QRS duration.

3.2 S3 Heart Sound

Table 2 shows the importance of considering the duration of the ECG QRS complex when using electronically recorded heart sound data to detect LVSD. Table 2 reveals that electronically recorded heart sounds were recorded from 432 patients and of this total number of patients, 107 had ECG QRS durations >120 ms. In the entire group, the prevalence of LVSD was 49.5% and in the subgroup with prolonged QRS duration, the prevalence of LVSD was 79.4%. In the entire group of patients, the diagnostic sensitivity of the electronically recorded S3 for detecting LVSD at 98% specificity was 21.0%. In the subgroup with prolonged QRS duration, the diagnostic sensitivity of the recorded S3 for detecting LVSD was 31.8% and this improvement in diagnostic performance reached statistical significance. Table 2 also shows that in the entire group of 432 patients compared to the subgroup with prolonged QRS duration, the thresholds of the recorded S3 display values needed to attain 98% specificity were 5.66 and 4.99, respectively. In addition, the ratio of true positive to false positive S3 test results was 10.3 in the entire group and rose to 27.0 in the subgroup with prolonged QRS duration.

As expected, the prevalence data in the third columns of both Tables 1 and 2 show that the prevalence of LVSD is greater in the subgroup of patients with QRS duration >120 ms. than it is in the entire population

Table 1
BNP Data for Detecting LVSD

Group	N	Prevalence of LVSD	Sensitivity @ 98% Specificity	Chi Square	P Value	Threshold Value*	TP/FP Ratio
All Patients	374	48.9%	10.9%			1740	5.2
QRS >120	183	82.8%	46.8%	41.2	$1.4x10^{-10}$	407	112.5

*in pg./ml.

BNP = blood natriuretic peptide, FP = false positive, LVSD = left ventricular systolic dysfunction, QRS = ECG QRS complex duration (ms.), TP = true positive

Table 2
S3 Data for Detecting LVSD

Group	N	Prevalence of LVSD	Sensitivity @ 98% Specificity	Chi Square	P Value	Threshold Value*	TP/FP Ratio
All Patients	432	49.5%	21.0%			5.66	10.3
QRS >120	107	79.4%	31.8%	3.8	0.05	4.99	27.0

*in proprietary recorded heart sound display values

FP = false positive, LVSD = left ventricular systolic dysfunction, QRS = ECG QRS complex duration (ms.), TP = true positive

Table 3
Comparison of BNP, S3 and QRS Duration Data as Parameters For Detecting LVSD

Parameter	Group	LVSD Present	Sensitivity @ 98% Specificity	Threshold Value	Chi Square[1]	P Value[1]
QRS Duration[2]	All	215	20.0	151		
BNP[3]	All	183	10.9	1740	6.11	0.01
BNP[3]	QRS>120	77	46.8	407	20.6	$5.7x10^{-6}$
S3 DV	All	214	21.0	5.66	.07	NS
S3 DV	QRS>120	85	31.8	4.99	4.71	0.03

[1]compared to QRS Duration, [2]in ms., [3]in pg./ml.

BNP = blood natriuretic peptide, DV = proprietary display values, LVSD = left ventricular systolic dysfunction, S3 = third heart sound

of patients. In other words, the presence of QRS duration >120 ms. increases the prior probability that LVSD is present. In keeping with the principles of Bayesian statistics, these higher prior probabilities of LVSD are associated with improved diagnostic performances of both the BNP and the recorded heart sound tests.[6]

The observed association of prolonged QRS duration with improved diagnostic performances of both BNP and recorded heart sounds raises the question of whether QRS duration itself is a useful parameter for detecting LVSD and the data shown in Table 3 address this possibility. Table 3 indicates that in the entire population of patients, the diagnostic sensitivity at 98% specificity of

QRS duration is significantly better (20.0% vs. 10.9%) than that of the BNP test when the latter is used in the lower prevalence total population. To reach 98% specificity for LVSD, the QRS duration must be markedly prolonged to 151 ms. However, in patients whose QRS durations exceed 120 ms., the sensitivity of the BNP test highly significantly exceeds that of both QRS duration alone and the BNP test applied without regard to QRS duration. Table 3 also shows that QRS duration and the S3 heart sound test have similar diagnostic performances when both parameters are used in the entire population of patients. However, when applied to patients whose QRS duration exceeds 120 ms. the performance of the S3 heart sound test is significantly better than that of the QRS duration. Thus, for both the BNP blood test and the S3 heart sound test for LVSD, the best diagnostic results are obtained when the ECG QRS duration has been used to identify subgroups that have higher prevalences of LVSD than those of the entire population.

4.0 Conclusions

The results of the present study support the hypothesis that a precise and readily available ECG parameter – the duration in ms. of the QRS complex – can be used to identify subgroups of patients in which the prevalence of LVSD is higher than the prevalence of LVSD in the entire population of patients. This is particularly remarkable because in contrast to an asymptomatic "screening" population; the prevalence of LVSD in adults reporting to emergency departments with shortness of breath is already high. Thus, the use of the QRS duration to identify subgroups with a higher prevalence of LVSD has incremental value over using just the clinical circumstances of the population being evaluated. The findings of the present study also support the hypothesis that the results of tests specifically intended to detect LVSD exhibit better diagnostic performances in the subpopulations with higher prevalences of LVSD than in the populations that have lower prevalences of this abnormality. Tables 1 and 2 show these results for the BNP data and for the S3 heart sound data, respectively. A way to emphasize the importance of these findings is from the perspective of considering the meaning of a "positive" result of a test for LVSD in a given patient. The last column of Table 1 shows that for the BNP test, the ratio of true positive to false positive results is 5.2 in the lower prevalence total group compared to 112.5 in the higher prevalence subgroup. This means that a given positive result on the BNP test in the higher prevalence subgroup compared is 112.5/5.2 = 21.6 times more likely to be correct than it is in the lower prevalence total group. The last column of Table 2 shows that a given positive result on the S3 heart sound test in the higher prevalence subgroup is 27.0/10.3 = 2.6 times more likely to be correct than it is in the lower prevalence total group.

The data in Table 3 indicate that using ECG QRS complex duration in ms. itself can be used as a parameter for detecting LVSD, although the diagnostic sensitivity at 98% specificity is only 20.0%. Nevertheless, when applied to the lower prevalence populations, the diagnostic sensitivity of QRS duration alone is better than the BNP test (20.0% vs. 10.9%) and equivalent to the S3 heart sound test in the low prevalence population (20.0% vs. 21.0%). However, in the high LVSD prevalence subpopulations, using QRS duration alone is diagnostically inferior to both BNP and the S3 heart sounds (sensitivities at 98% specificity of 46.8% and 31.8%, respectively). These data show that for detecting LVSD, using the ECG QRS duration to identify subgroups with high prior probabilities of LVSD is superior to using the QRS duration as a standalone diagnostic test.

A previous study from this laboratory showed that one may also assess the prior probability of LVSD using ECG evidence of prior myocardial infarction.[7] However, the use of the QRS duration on the ECG has important advantages. First, there are many different sets of diagnostic criteria for myocardial infarction and the accuracy of each of them is imperfect. Second, the presence or absence of myocardial infarction is irrelevant in the many patients whose LVSD has been caused by pathology other than coronary artery disease. In contrast, the measurement of the QRS complex is ubiquitous, extremely accurate and highly reproducible. QRS duration in ms. is routinely reported in both computerized and visual ECG interpretations. Additional work will be required to determine whether ECG findings other than criteria for myocardial infarction or measurements of QRS duration are superior for assessing the prevalence of LVSD in a given population.

In everyday life, we often informally consider prior probability when trying to decide whether something is true or false. This involves thinking about the background circumstances in which the truth or falsity of an assertion is being evaluated. For example, we tend to be less likely to accept uncritically a salesman's description of a product if we know that he is trying to sell that product to us. Prior probability is also considered in medical diagnosis, even though such considerations are also often informal. For example, if a particular disease is especially common in elderly women, a physician knows that a positive result on a diagnostic test for that disease is less likely to be correct if the test had been performed on a young male. The present study shows that computerized ECG data can be used in a formal, quantitative way to affect the prior probability that a certain disease is present. This in turn can significantly improve the diagnostic performances of non-ECG tests for that disease. The specific findings of the present study are especially important because LVSD can result from a wide variety of types of underlying heart disease (e.g. coronary artery disease, hypertension, infection, valvular heart disease, cardiomyopathy and congenital heart disease) and it often leads to premature disability and death. Therefore, improvements in the accuracy of diagnostic tests for LVSD can increase the likelihood that patients receive timely and effective treatment for this important condition.

5.0 References

1. Cleland JG. Long-term mortality with cardiac resynchronization therapy. Eur. J. of Heart Failure. 14:628-634, 2012
2. Kumar, Vinay; Abbas, Abul K; Aster, Jon. (2009). Robbins and Cotran pathologic basis of disease (8th ed.). St. Louis, Mo: Elsevier Saunders. p. 574. ISBN 1-4160-3121-9.
3. Seino Y, Ogawa A, Yamashita T, et al. Application of NT-proBNP and BNP measurements in cardiac care: a more discerning marker for the detection and evaluation of heart failure. Eur J Heart Fail. Mar 15. 6(3):295-300, 2004
4. Kuo PT, Schnabel TG, Blakemore WS, Whereat AF. Diastolic gallop sounds, the mechanism of production. J Clin. Invest. 36 (7): 1035–42, 1957
5. Warner RA, Anderson E and Arand P. Enhancement of the detection of left ventricular enlargement by the combination of ECG and acoustical findings. International Journal of Bioelectromagnetism. 5(1):193-196, 2003.
6. Warner RA. Optimizing the Display and Interpretation of Data. Amsterdam, Elsevier, 117-134, 2015.
7. Warner RA. Using the principles of bayesian statistics to improve the performances of medical diagnostic tests. Proceedings of the 2014 International Conference on Computa-

tional Science and Computational Intelligence Edited by B. Akhgar and HR Arabnia, IEEE Computer Society CPS, USA, 2014, P. 64-68

Research on Classification of Diseases of Clinical Imbalanced Data in Traditional Chinese Medicine

Zhu-Qiang Pan
School of Computer Science, Southwest Petroleum University
Chengdu 610500, China
panzhuqiang@foxmail.com

Mary Qu Yang
MidSouth Bioinformatics Center
University of Arkansas Little Rock College of Engineering & IT and University of Arkansas for Medical Sciences.
2801 S. University Avenue, Little Rock, Arkansas 72204
U.S.A
maryy9505@gmail.com

Lin Zhang
School of Computer Science, Southwest Petroleum University
Chengdu 610500, China
linzhang8080@163.com

Guo-Zheng Li
China Academy of Chinese Medical Science
Beijing 100700, China
gzli@ndctcm.cn

***Abstract*—Traditional Chinese medicine (TCM) on certain diseases are likely to be unbalanced, and this unbalanced data tends to be biased towards disease-free individuals. In view of this problem, this paper proposes an FPUSAB algorithm to deal with the problem of unbalanced classification of clinical disease data in TCM with improved under-sampling. Experimental results on the meridian resistance data collected by traditional Chinese medicine show that the FPUSAB algorithm improves the classification performance.**

***Keywords*—Chinese medicine clinical; disease; imbalance data classification**

I. Introduction

Data mining is becoming more and more important in Traditional Chinese medicine (TCM) diagnosis, and computer-aided diagnosis is essentially a data mining classification task [1]. The classification performance directly affects the ability of auxiliary diagnosis. In real world, a lot of data is not balanced. For example, in the medical diagnosis, individuals suffering from a disease are often minority; mechanical fault detection[2] studies have shown that in the rotating machinery gear failure accounted for about 10% of its failure. Similar problems exist in the field of image detection, communication field customer loss prediction[3]and other fields. For the classification of unbalanced data, the traditional data mining classification methods tend to negative (more a class of data), and for positive (less a class of data) classification is poor. But in real life, people pay more attention to positive. For example, in the process of disease classification of TCM clinical data, researchers pay more attention to the classification of diseased individuals. Positive classification performance direct impact on the computer's diagnostic capabilities, but also related to the doctor's diagnostic efficiency. In the classification of imbalanced data, the expense of the positive classes is much higher than the expense of negative classes, and some of the traditional methods of "preference" negative are no longer applicable.

Imbalanced data has attracted researcher's attention. In recent years, many algorithm is proposed. In view of the unbalanced data classification of the existing algorithms mainly from the data set, classifier, classifier and data set of these three ways[4]to deal with the imbalanced data classification. From the data set is mainly under- sampling and over-sampling, but these two methods are not reveal the actual characteristics of the data, so the classification performance needs to be further improved. In clinical imbalanced data, if only use the under-sampling, may lost a lot of important information of the original data; over-sampling simple copy the positive data will appear over-fitting phenomena.

In this paper, an improved algorithm FPUSAB is proposed to deal with the problem of unbalanced classification by combining the actual situation of TCM unbalanced data, combined with under-sampling and Asymmetric Bagging[5].

II. Measures

Since the class distribution of the data set is unbalanced, only correction of classification accuracy may be misleading. Therefore, AUC (Area Under the Curve of Receiver Operating Characteristic (ROC)) [6] is used to measure the performance.

At the same time, in view of the shortcomings of the traditional classification performance, many scholars in the study of imbalance data classification using the following performance measures. Table I for the two classes of confusion matrix, *TP*, *FP*, *FN*, *TN*, respectively, on behalf of the number of true negative, false positive, false negative, true negative.

Table I confusion matrix

	Predict positive	Predict negative
Real positive	*TP*	*FN*
Real negative	*FP*	*TN*

Sensitivity is defined as:

$$Sensitivity = \frac{TP}{TP + FN} \quad (1)$$

Specificity is defined as:

$$Specificity = \frac{TN}{TN + FP} \quad (2)$$

Bacc((Balanced Accuracy)) is defined as :

$$Bacc = \frac{1}{2}(\frac{TP}{TP+TN} + \frac{TN}{TN+FP}) \quad (3)$$

PPV(Positive Predictive Value) is defined as :

$$ppv = \frac{TP}{TP+FP} \quad (4)$$

NPV(Negative Predictive Value) is defined as :

$$npv = \frac{TN}{TN+FP} \quad (5)$$

Correction((Balanced Accuracy)) is defined as:

$$Correction = \frac{TP+TN}{TP+TN+FP+FN} \quad (6)$$

III. Data Level Solving Unbalanced Classification Method

From the data level, in the process of reconstructing the data set, a mechanism is used to obtain a more balanced data distribution, which is called resampling, equivalent to a preprocessing data equalization method. Researchers have proposed a variety of sampling techniques, it can be divided into three kinds: under-sampling, over-sampling, based on the former two mixed sampling [7].

Under-sampling refers to the removal of some samples from the original data set to achieve the same number of samples in the class. The most commonly used is random under-sampling[8],it randomly remove negative samples from the original data set, reducing the size of negative to achieve the more balanced data set. However, this method may lose the representative of the majority of samples information when eliminate the majority of samples, resulting in loss of information affect the classification effect. Unlike under-sampling, over-sampling[9] is the use a mechanism to add samples to the original dataset, making the negative and positive balanced. The most commonly used is random over-sampling, it randomly copying positive samples to make the data balance distributed. Since random over-sampling just simply adds positive of copies to the original dataset, there will be a lot of "duplicate" samples, resulting in over-fitting [10].

Zhao et al [11] pointed out the advantages and disadvantages of under-sampling and over-sampling and proposed a new sampling method based on under-sampling and achieved a better result. However, this sampling method is mainly to get balance as close as possible, not fundamentally solve the problem of imbalance. At the same time, for existing sampling methods, existing research attempts to combine under-sampling and over-sampling. For example, Zhu et al[12]proposed the RU-SMOTE-SVM algorithm, which combines the random under-sampling method and the SMOTE algorithm for artificially synthesizing positive samples. Li et al[13]combined with the mixed sampling strategy and Bagging proposed Asymmetric Bagging(AB) algorithm, AB has achieved a better result in the bioinformatics imbalanced data classification.

TCM clinical data is collected from the patient's physical signs related to the actual data, due to question the authenticity of the synthesis, so the clinical data of TCM less use SMOTE artificial synthesis of positive samples to deal with the disease classification. Simultaneous use over-sampling randomly selected samples of the original positive, copy and add to the original set is also very easy to cause over-fitting. But for under-sampling and over-sampling, Drummond[14]et al believe that under-sampling is superior to over-sampling in performance.

IV. FPUSAB Algorithm

In TCM clinical data, each sample is an individual vital signs data, and when we put them into the sample space, each sample is a sample point of the sample space[15]. In the case of random under-sampling, if a sample point in a finite area is retained, there may be a large number of valuable samples points discarded; if the randomly selected samples are concentrated in a certain area, will cause the phenomenon of over-fitting. The Corresponding to the actual situation: If we select a number of patients with the same characteristics and not sick in the selection of patient cases, then according to their situation to determine other people who do not have these characteristics of the situation, often do not get the results, or judgments tend to be random. If a certain amount of samples are retained in each area of the samples, the worst "distortion" condition can be prevented. For a region sample, they should be at a fixed distance. Corresponding clinical practice: in a similar characteristics of the patient group selected one stand for this group, and each group selected one,then encounter a new patient, more of the judge basis ,it can be more effective on the disease classification.

Therefore, in order to maintain the majority of the sample's original information characteristics in an under-sampling process, the following approach is proposed[11]. The black dots in Figure 1 (a) are the mean points of the majority of samples. Calculate the distance between all the negative samples and the mean points. In each small area where the distance is close, a point is left and remove the remaining points. All of the selected negative samples remain together as a new negative samples set and the original positive samples together to form a new training set, as shown in Figure 1 (b).

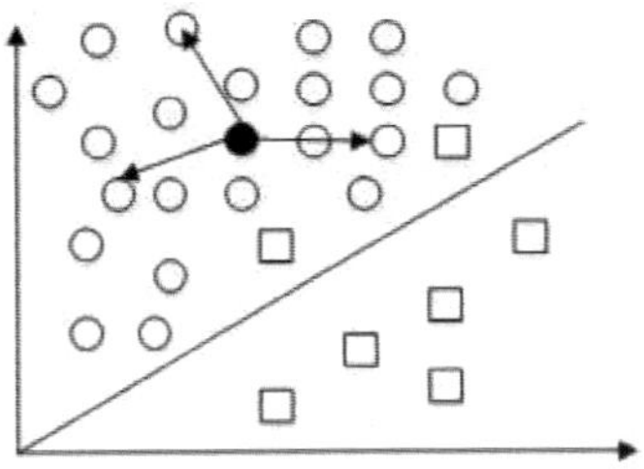

(a)

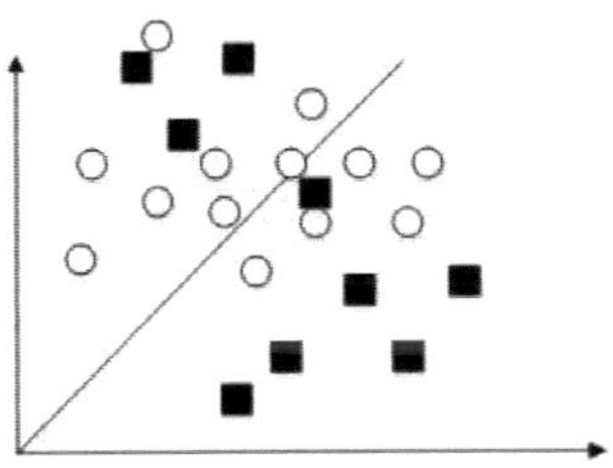

(b)

Figure 1 Furthest patient

The traditional classification algorithm has a good performance on the balanced data set. Asymmetric Bagging algorithm based on the idea of balanced and random under-sampling, each from negative samples randomly selected with a small number of equal positive samples, and then this part of the samples and positive together to form a new data set, and then repeated this process to form multiple training subsets, then Asymmetric Bagging will train the training subsets by SVM, the final classification results determined by the obtained models. Due to random under-sampling, It can not avoid appearing the "distortion".

Input： Training data set S_r={(x,y)},number of individuals T

Output： Ensemble model N

1： begin

2： **for** k=1： T **do**

3： Generate a training subset S_{rk}^- from negative set S^- using bootstrap sampling technique , the size of S_{rk}^- is the same as with that of S^+.

4： Train the individual model N_k on the training subset $S_{rk}^- \cup S^+$ by using support vector machines.

5： **end for**

6： Ensemble the obtained models N by the way of majority voting method for classification problems

7： end

Figure 2 Asymmetric Bagging algorithm

Input： Training data set S_k, number of bagsize N

Output： Ensemble model N

Method： FPUSAB approach

1： begin

2： **for** k=1： T **do**

3： Generate a training subset S_k^- from negative training subset S^- and count the number $count_{ne}$.

4： Generate a training subset S_k^+ from positive training subset S^+ and count the number $count_{po}$.

5： compute the mean value of the sample in Training subset S_k^-： meanOfNegative, sort the sample from according to the distance from large to small and stored in the S_{stored}^-.

6： if(N>($count_{ne}$% $count_{po}$))break;

7： **for** i=1： N **do**

8： **for** j=1： $count_{po}$ **do**

9： set count =0;

10： if(>(count% N)==0) remove the $sample_j$ from the S_{stored}^- to S_{new}^-;

count++;

11： **end for**

12： $S_{new}^- \cup S_k^+$ constitute the subset S_{ti}^-.

13： Train an individual model N_i on the training subset S_{ti}^- by using support vector machines.

14： **end for**

15： Ensemble the obtained models N by the way of majority voting method for classification problems.

16： end

Figure 3 FPUSAB algorithm

As shown in Figure 3, in the FPUSAB(Furthest Patient based on Under Sampling for Asymmetric Bagging) algorithm, First ,calculate the distance between the each negative sample and the center point (the negative samples mean points), and sort the negative samples according to the distance from large to small to form the *M*. And then according to the number of bags in the Bagging (the number of ensemble models) to select a small number of samples from *M* to constitute a number of training subset, these subsets are trained by SVM to form models. Finally, the results of the classification of the testing set are determined by these small models voting.

V. DATA SET

The experiment data derived from the clinical collection of TCM clinical meridian resistance data. Among the 3053 samples collected, the data of the different classes were different. After deleting the severely missing data and filling the data with not severe data, we found 534 samples of health and sub-health, and 439 samples of health data and 95 samples of sub-health data. For the remainder of sleep disorders, 2214 samples, sleep disorders include three sub-types: specific sleep disorders, anxiety disorders, depression. Before the experiment, we have made some merges for the sample class of the dataset, all merged into two classes classification problems. Then we found suffering from sleep disorders 206 samples, not suffering 2008 samples.

For the collection of TCM clinical data can be found the health of individuals over sub-health individuals, in the number of patients with sleep disorders are not more than the number of not sick individuals. It should be noted that in the traditional Chinese medicine, It does not contain sub-health of the disease and sleep emotional disease. Sub-health, sleep disorders are Western medicine diagnosis, this paper's research combined with the clinical data of traditional Chinese medicine for Western medicine disease classification. In Table II, health indicates sub-health disease, sleep indicates sleep disorder, and Ratio represents the ratio of the negative to the positive.

Table II experiment dataset

Disease	class	Feature	size	Min/ Max	Ratio
health	2	28	534	95/435	4.57
sleep	2	28	2214	206/2008	9.74

VI. EXPERIMENTS AND REESULTS

In order to analyze the performance of the algorithm, a variety of methods for experimental analysis. In the traditional classification algorithm, we choose decision tree(J48)[16]、Naive Bayes[17] 、 SVM[18] 、 Bagging, In the existing unbalanced data classification algorithm, select the unbalanced support vector machine (unSVM[19]), Bagging based on unSVM unbalanced Bagging (unBagging[19]) and Asymmetric Bagging algorithm. Compare with FPUSAB and the above seven methods. All experiments were performed using 10-fold cross validation to assess AUC and related properties. To exclude randomness, Each experiment was repeated 10 times. decision tree (J48), Naive Bayes, Bagging, using JAVA language implemented in Weka [20]; SVM, unSVM, unBagging, Asymmetric Bagging using JAVA language implemented in LibSVM . Related programming are based on JAVA language. In order to facilitate comparison, Bagging, Asymmetric Bagging, FPUSAB, SVM use the same parameter settings, in the experiment,the parameters used the default parameters and the ensemble scale set to 1.

Table III health that sub-health diseases, Table IV sleep that sleep disorders, the table in the unit %. In the table will be Asymmetric Bagging abbreviated as AB, the best evaluation of the indicators marked in bold.

Table III Chinese medicine clinical health imbalance data classification results

disease	method	*AUC*	*Sensitivity*	*Specificity*	*Bacc*	*ppv*	*npv*	*Correction*
health	J48	50.4	7.4	**95.9**	51.7	25.1	83.0	80.1
health	Naive Bayes	66.3	29.5	83.1	56.3	27.5	84.5	74.2
health	SVM	50.0	10.0	94.0	52.0	15.0	**90.7**	82.0
health	unSVM	52.0	12.0	92.0	52.0	15.2	92.0	83.0
health	Bagging	54.7	11.6	85.9	48.8	15.1	81.8	72.7
health	unBagging	55.0	10.0	86.0	48.0	15.3	82.4	73.1
health	AB	66.7	**73.7**	51.7	62.7	25.0	90.0	55.7
health	FPUSAB	**71.7**	63.2	70.1	**66.65**	**31.5**	89.7	68.9

Table IV Chinese medicine clinical sleep disorders disease imbalance data disease classification results

disease	method	*AUC*	*Sensitivity*	*Specificity*	*Bacc*	*ppv*	*npv*	*Correction*
sleep	J48	52.8	14.1	94.5	50	10	90.7	82.3
sleep	Naive Bayes	69.2	18	**95.7**	56.85	**22.5**	90.6	86.3

sleep	SVM	50	15	92	50	20	90.7	81
sleep	unSVM	51	16	93	54.5	21	90.2	83
sleep	Bagging	55.6	6.8	95	50	13.3	90	**86**
sleep	unBagging	56.1	7.1	94.5	50.8	13.5	89.6	85.4
sleep	AB	65.6	60.9	58.8	59.85	14.3	93	68.4
sleep	FPUSAB	**70.4**	**65.8**	62.1	**63.95**	16.4	**94.1**	62.5

From the Table III, Table IV can be found, for imbalanced data classification ,the traditional classification algorithm decision tree (J48), Naive Bayes, SVM has a poor performance; AB, FPUSAB has a better performance; unSVM does not effectively improve the performance of SVM, unBagging compared to Bagging is only a small improvement in performance; Bagging also poor. The FPUSAB algorithm is superior to other algorithms for the main classification indicators *AUC* and *Bacc*.

What kind of impact of the number of bags (ensemble scale) on the classification? If the bags increasing, Asymmetric Bagging algorithm will be better than FPUSAB algorithm? We continue to experiment to validate. Due to health, sub-health and sleep disorders are not equal at Ratio, so the number of bags is also different. According to the Ratio, we limit the number of health bags to 4, the number of sleep bags to 9. Due to the classification performance mainly determined by *AUC*, *Bacc*, so in the latter these two measures were analyzed.

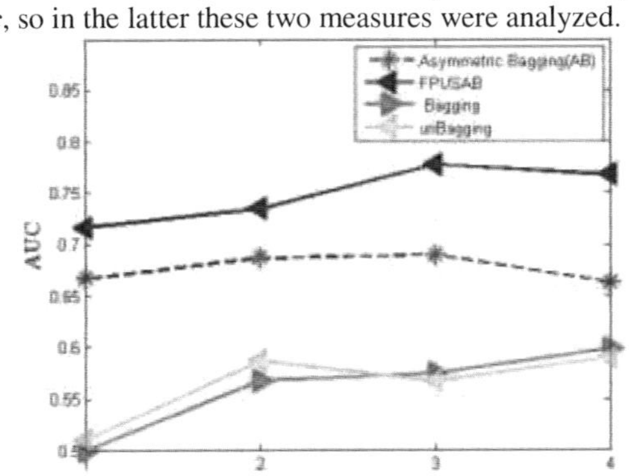

(a) sub-health disease AUC results

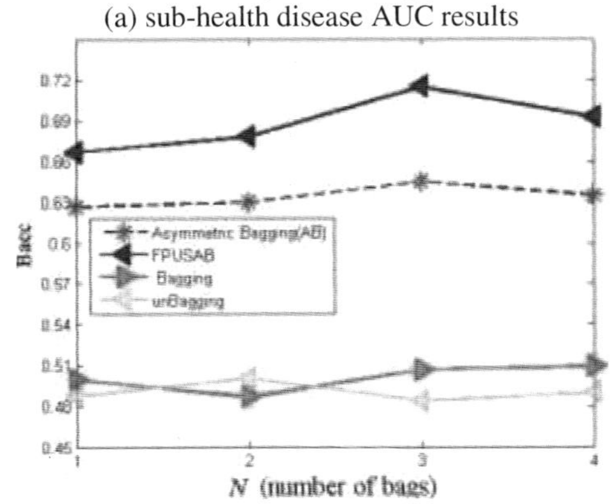

(b) sub-health disease Bacc results

Figure 4 sub-health disease classification results

As can be seen from Figure 4, with the increase of the bags, *AUC*, *Bacc* appears to increasing. As a result of the random under-sampling, Bagging, unBagging with the increase of the bags changes in oscillation and worse than Asymmetric Bagging and FPUSAB. For Asymmetric Bagging, FPUSAB appeared a relatively better increasing; and on the whole FPUSAB is better than Asymmetric Bagging. We can found when *N* is greater than 3, Asymmetric Bagging in the classification performance of the decline is greater than FPUSAB, indicating that FPUSAB's stability is better than Asymmetric Bagging. When *N* is 3, FPUSAB, Asymmetric Bagging works best. For the best *AUC* , FPUSAB algorithm is about 0.77, Asymmetric Bagging algorithm is about 0.67. For the best *Bacc* , FPUSAB algorithm *Bacc* is about 0.71, Asymmetric Bagging algorithm *Bacc* is about 0.63. On the whole, FPUSAB is better than Asymmetric Bagging.

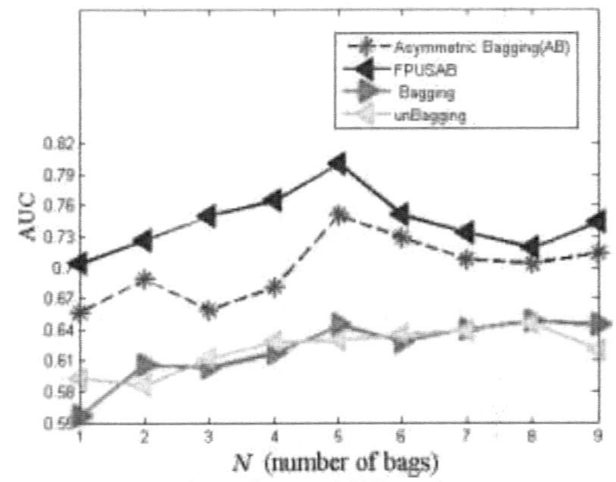

(a) sleep disease AUC results

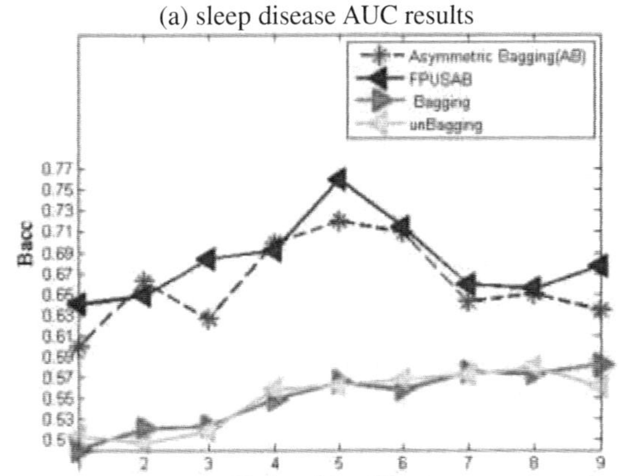

(b) sleep disease Bacc results

Figure 5 sleep disease classification results

As can be seen from Figure 5, the *AUC*, *Bacc* has different change with the number of bags and show different trends. On the whole， Bagging, unBagging is increasing classification performance, but the increasing is not significant and worse than Asymmetric Bagging, FPUSAB. For Asymmetric Bagging, FPUSAB, when the *N* is less than 5, Asymmetric Bagging has a oscillation increase, and FPUSAB has a more stable growth and in the classification performance FPUSAB better than Asymmetric Bagging ; when *N* is more than 5, Asymmetric Bagging, FPUSAB has a declining trend, from the range of decline, FPUSAB is better than Asymmetric Bagging. When *N* is 5, FPUSAB, Asymmetric Bagging works best. For the best *AUC* ,FPUSAB algorithm is about 0.80, Asymmetric Bagging algorithm is about 0.75. For the best *Bacc* ,FPUSAB algorithm *Bacc* is about 0.77, Asymmetric Bagging algorithm *Bacc* is about 0.71. On the whole, FPUSAB is better than Asymmetric Bagging.

From Figure 4, Figure 5 can be found, for the classification the sleep disorders is superior to sub-health. The main reason is the unbalanced degree of sleep emotional diseases (Ratio 9.74) more than the sub-health diseases (Ratio 4.57). From here we can see that FPUSAB is more effective for the clinical imbalance of higher data. we also found that the size of the optimal effect ensemble scale based on under-sampling is about half that of the unbalanced scale. For example, the best scale for sub-health when *N* is 3, the best for sleep when *N* is 5.

Compared with the Asymmetric Bagging, for the classification of health diseases, FPUSAB algorithm has an average increase of 12.7% on the *AUC* and 10.8% on the *Bacc*;For the sleep disease classification, the FPUSAB algorithm averaged increase 7.4% on the *AUC* and 6.2% on the *Bacc*. In general, the FPUSAB algorithm averaged 10.5% on the *AUC* and 8.4% on *Bacc*.

In a word, FPUSAB algorithm is better than Bagging, unBagging, Asymmetric Bagging. Compared with the Asymmetric Bagging algorithm, the FPUSAB algorithm improves the classification performance.

VII. CONCLUSIONS

In order to improve the classification performance of TCM clinical unbalanced data, an improved algorithm FPUSAB of Asymmetric Bagging was proposed in combination with improved under-sampling. Experiments were carried out to collect clinical data of TCM, and compared with the traditional classification algorithm and the existing unbalanced data classification algorithm. The experimental results show that compared with the Asymmetric Bagging algorithm, the FPUSAB algorithm is an average of 10.5% on the *AUC* and 8.4% on the *Bacc*. In the existing unbalanced data classification algorithm, FPUSAB has the best classification effect and better stability.

Although this work improves the classification performance of TCM unbalanced data, there is still much work to be done, such as further improving the sampling method and making the classification more better.

REFERENCES

[1] Y. Zou, "APPLYING FEATURE SELECTION-BASED CLASSIFICATION ENSEMBLE IN SPLEEN ASTHENIA DIAGNOSIS," *Computer Applications & Software,* 2010.

[2] T. Y. Liu, "Research on imbalanced problems in gear fault diagnosis," *Computer Engineering & Applications,* 2006.

[3] N. Xie, B. Fang, and W. U. Lei, "Study of text categorization on imbalanced data," *Computer Engineering & Applications,* 2013.

[4] T. Y. Liu and L. I. Guo-Zheng, "The Imbalanced Data Problem in the Fault Diagnosis of Rolling Bearing," *Computer Engineering & Science,* 2010.

[5] D. Tao, X. Tang, X. Li, and X. Wu, "Asymmetric bagging and random subspace for support vector machines-based relevance feedback in image retrieval," *IEEE Transactions on Pattern Analysis & Machine Intelligence,* vol. 28, pp. 1088-99, 2006.

[6] J. H. Xue and P. Hall, "Why Does Rebalancing Class-Unbalanced Data Improve AUC for Linear Discriminant Analysis?," *IEEE Transactions on Pattern Analysis & Machine Intelligence,* vol. 37, pp. 1109-1112, 2015.

[7] X. Tao, S. Hao, D. Zhang, and X. U. Peng, "Overview of classification algorithms for unbalanced data," *Journal of Chongqing University of Posts & Telecommunications,* vol. 25, pp. 101-43, 2013.

[8] M. A. Tahir, J. Kittler, and F. Yan, "Inverse random under sampling for class imbalance problem and its application to multi-label classification," *Pattern Recognition,* vol. 45, pp. 3738-3750, 2012.

[9] M. J. Kim, D. K. Kang, and B. K. Hong, "Geometric mean based boosting algorithm with over-sampling to resolve data imbalance problem for bankruptcy prediction," *Expert Systems with Applications,* vol. 42, pp. 1074-1082, 2015.

[10] J. Pan and L. I. Hong, "Research on classification algorithms in imbalanced data based on boosting," *Computer Engineering & Applications,* vol. 45, pp. 138-140, 2009.

[11] Z. Zhao, G. Wang, and L. I. Xiaodong, "An Improved SVM Based Under-Sampling Method for Classifying Imbalanced Data," *Zhongshan Daxue Xuebao/acta Scientiarum Natralium Universitatis Sunyatseni,* vol. 51, pp. 10-16, 2012.

[12] X. M. Tao, Z. J. Tong, Y. Liu, and D. D. Fu, "SVM classifier for unbalanced data based on combination of ODR and BSMOTE," *Kongzhi Yu Juece/control & Decision,* vol. 26, pp. 1535-1541, 2011.

[13] H. H. Meng, M. Q. Yang, and J. Y. Yang, "Asymmetric Bagging and Feature Selection for Activities Prediction of Drug Molecules," in *International Multi-Symposiums on Computer and Computational Sciences*, 2007, pp. 108-114.

[14] C. Drummond and R. C. Holte, "C4.5, Class Imbalance, and Cost Sensitivity: Why Under-Sampling beats OverSampling," *Proc of the Icml Workshop on Learning from Imbalanced Datasets II,* pp. 1--8, 2003.

[15] X. Fei, X. Li, and C. Shen, "Parallelized text classification algorithm for processing large scale TCM clinical data with MapReduce," in *IEEE International Conference on Information and Automation*, 2015, pp. 1983-1986.

[16] D. N. Bhargava, G. Sharma, R. Bhargava, and M. Mathuria, "Decision tree analysis on j48 algorithm for data mining," 2013.

[17] J. Salvador and E. Perezpellitero, "Naive Bayes Super-Resolution Forest," in *IEEE International Conference on Computer Vision*, 2015, pp. 325-333.

[18] Y. Bazi and F. Melgani, "Toward an Optimal SVM Classification System for Hyperspectral Remote Sensing Images," *IEEE Transactions on Geoscience & Remote Sensing,* vol. 44, pp. 3374-3385, 2006.

[19] C. W. Hsu, C. C. Chang, and C. J. Lin, "A Practical Guide to Support Vector Classification Chih-Wei Hsu, Chih-Chung Chang, and Chih-Jen Lin," 2003.

[20] I. H. Witten and E. Frank, "Data mining: practical machine learning tools and techniques with Java implementations," *Acm Sigmod Record,* vol. 31, pp. 76-77, 2011.

Fluorescence Microscopy Noise Model: Estimation of Poisson Noise Parameters from Snap-Shot Image

Y. Kalaidzidis[1,2]
[1] Max Planck Institute of Molecular Cell Biology and Genetics, 01307 Dresden, Saxony, Germany
[2] Faculty of Bioengineering and Bioinformatics, Moscow State University, 119991 Moscow, Russia

Abstract - *The estimation of image noise parameters is crucial for quantitative microscopy image analysis. Although in time-lapse sequences the noise estimation is straightforward (see below), it is not so for single image. In this work, I describe a simple method of Poisson noise estimation for single fluorescent microscopy images. The accuracy of the method was verified by comparing noise parameters learned from time-lapse and single snap-shot images.*

Keywords: Fluorescence microscopy; Poisson noise; BioImage analysis;

1 Introduction

Fluorescence microscopy is widely used in bio-medical research. Multiple algorithms were developed for quantitative microscopy image analysis [1-4]. The accuracy of the analysis outcome is crucially dependent on the proper estimation of the image noise parameters. Whereas simple algorithms for noise estimation in the time-lapse sequence relatively straightforward (e.g. [5]), published algorithms for single image noise estimation are more complicated [6,7]. Here, I propose a simple algorithm for estimation parameters of noise from single fluorescent microscopy images that circumvent the necessity of microscope gain calibration by independent experiments.

1.1 Noise model for fluorescent microscopy

The main source of the fluorescence microscopy noise is a shot noise of photons and photo-electrons in photomultiplier tube (PMT) or CMOS/CCD camera. The number of detected photo-electrons follows a Poisson probability distribution

$$P(N) = \frac{\lambda^N e^{-\lambda}}{\Gamma(N+1)} \qquad (1)$$

where N is a number of detected photo-electrons; λ is a mean number of photo-electrons. The signal from camera/PMT is amplified and digitalized to result in pixel intensity of the digital image. Therefore

$$I = c \cdot N + \beta \qquad (2)$$

where I is an intensity, c is gain coefficient of an amplifier and β is an image offset, i.e. the image intensity in the absence of light (β could be as negative as positive, depending on the microscope "dark current" settings).

Combining (1) and (2) we get the probability distribution for intensities:

$$P(I) = \frac{\lambda^{\frac{I-\beta}{c}} e^{-\lambda}}{\Gamma\left(\frac{I-\beta}{c}+1\right)} \frac{1}{c} \qquad (3)$$

The mean intensity from distribution (3) is:

$$\langle I \rangle = \sum_{\frac{I-\beta}{c}=0}^{\infty} I \cdot \frac{\lambda^{\frac{I-\beta}{c}} e^{-\lambda}}{\Gamma\left(\frac{I-\beta}{c}+1\right)} \frac{1}{c} = c \cdot \lambda + \beta \qquad (4)$$

Here, as one could expect, the mean intensity is the product of gain coefficient and mean number of photo-electrons plus offset. The Gaussian noise of amplifier is additive and has zero mean, therefore, it does not contribute to the mean intensity. Given that Poisson noise of the photo-electron flux is independent of Gaussian noise of amplifier, the variance of the intensity is sum of variance of photo-electrons counting and squared noise of amplifier:

$$\mathrm{var}(I) = \left(\sum_{\frac{I-\beta}{c}=0}^{\infty} I^2 \cdot \frac{\lambda^{\frac{I-\beta}{c}} e^{-\lambda}}{\Gamma\left(\frac{I-\beta}{c}+1\right)} \frac{1}{c} - \langle I \rangle^2 \right) + \sigma_a^2 \quad (5)$$

$$= c^2 \cdot \lambda + \sigma_a^2$$

where σ_a^2 is squared amplitude of amplifier noise. Combining (4) and (5) we get:

$$\mathrm{var}(I) = c \cdot \langle I \rangle + \varepsilon \quad (6)$$

where $\varepsilon = \sigma_a^2 - c \cdot \beta$. Note, that ε can be as positive as negative dependent sign and value of β.

1.2 Estimation of noise model parameters from time-lapse sequence

In order to determine c and ε, one can experimentally estimate variance of intensity in each pixel of image, average estimated variances of pixels of equal intensity and draw the dependency of $\mathrm{var}(I)$ as function of intensity I. One obvious way to do such estimation is to acquire the time-lapse sequence of the images of the sample, which mostly does not change between two sequential frames either because of high frame rate or because sample is fixed.

Then the variance could be estimated as a mean squared difference of pixel intensity between two sequential frames. The result of such procedure is presented on Fig.1. It is clear that the experimental data are well described by model (6) and parameters c and ε can be easily found by maximum likelihood fitting of strait line to the experimental data (Fig.1 solid line).

This method allows calibrate noise model parameters with specific microscope settings in advance and later use them for single snap-shot image analysis [Rink et al, 2005].

2 Estimation of Poisson noise model parameters by spatial variation

However, in many cases one has no access to time-lapse sequence and analysis is limited to single snap-shot image. Obviously, the time variance could not be simply substituted by space variance, since the last one is mixture of non-uniform intensity of the image and variance of noise. Fortunately, the variation of image intensity and noise fluctuations can be separated if pixel size is smaller than diffraction limit of microscope that is the case of high-resolution fluorescence microscopy. Non-uniform intensities of image have correlation length of Airy disk size (space diffraction limit), but Poisson fluctuations of neighbor pixels are not correlated. I use this feature of fluoresce high-resolution microscopy to estimate noise model from single image.

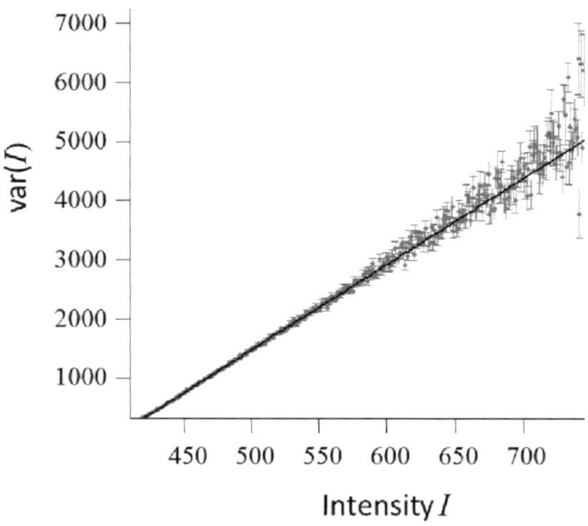

Figure 1. Variance of intensities estimated from fluctuations of pixel intensities in time-lapse sequence. GFP-Rab5c BAC HeLa cells were imaged by ANDOR Spinning Disc microscope (Nikon TiE inverted stand microscope equipped by spinning disc scan head CSU-X1; Yokogawa) with frame rate ~2.5 Hz. Gray dots are experimental estimation of variance of intensities (mean±SEM). Solid line is theoretical noise model (equation (6)).

2.1 Estimation of noise model parameters from single snap-shot image

Therefore, we convolved the image with Gaussian filter with kernel width equals to 0.5 pixel and subtract it from the original image. Since the diffraction-limited image was almost not disturbed by such mild filtering, the difference dominated by the noise. Therefore, we estimated variance for every pixel as a mean squared difference of intensity of differential image within 3x3 square neighborhoods. However, this procedure partially suppressed the noise. The noise suppression could be estimated as follow. Given that

$$\mathrm{var}(I - I_{smooth}) = \mathrm{var}(I) + \mathrm{var}(I_{smooth}) - 2 \cdot \mathrm{cov}(I_{smoothed}, I) \quad (7)$$

then, the variance of original image can be calculated as

$$\mathrm{var}(I) = \frac{1}{k} \mathrm{var}(I - I_{smooth}) \quad (8)$$

where

$$k = 1 + \frac{\mathrm{var}(I_{smooth})}{\mathrm{var}(I)} - 2 \cdot \frac{\mathrm{cov}(I_{smoothed}, I)}{\mathrm{var}(I)} \quad (9)$$

If the smoothing kernel width is much smaller than diffraction limit, then the variance of smoothed image could be approximated as:

$$\mathrm{var}(I_{smoothed}) \approx \mathrm{var}(I) \cdot \sum_{x,y} \left(\int_{x-0.5}^{x+0.5} \int_{y-0.5}^{y+0.5} \frac{1}{2\pi\sigma^2} e^{-\frac{1}{2}\frac{x^2+y^2}{\sigma^2}} dxdy \right)^2 \quad (10)$$

where σ is a width of filtering Gaussian kernel (we have chosen $\sigma = 0.5$).

The covariation of smoothed and original image is:

$$\mathrm{cov}(I_{smoothed}, I) = = \frac{N}{N-1}\left(\frac{1}{N}\sum_{x,y} I \cdot I_{smoothed} - \left(\frac{1}{N}\sum_{x,y} I\right)\left(\frac{1}{N}\sum_{x,y} I_{smoothed}\right)\right)$$

$$\approx \mathrm{var}(I) \cdot \sum_{x,y} \int_{x-0.5}^{x+0.5} \int_{y-0.5}^{y+0.5} \frac{1}{2\pi\sigma^2} e^{-\frac{1}{2}\frac{x^2+y^2}{\sigma^2}} dxdy \quad (11)$$

Therefore, substituting (10) and (11) into (9) we got expression for noise suppression factor:

$$k \approx 1 + \sum_{x,y} \left(\int_{x-0.5}^{x+0.5} \int_{y-0.5}^{y+0.5} \frac{1}{2\pi\sigma^2} e^{-\frac{1}{2}\frac{x^2+y^2}{\sigma^2}} dxdy \right)^2 - 2\sum_{x,y} \int_{x-0.5}^{x+0.5} \int_{y-0.5}^{y+0.5} \frac{1}{2\pi\sigma^2} e^{-\frac{1}{2}\frac{x^2+y^2}{\sigma^2}} dxdy \quad (12)$$

This correction factor is dependent only on width of smoothing kernel and can be calculated in advance (graph of $k(\sigma)$ is shown on Fig.2).

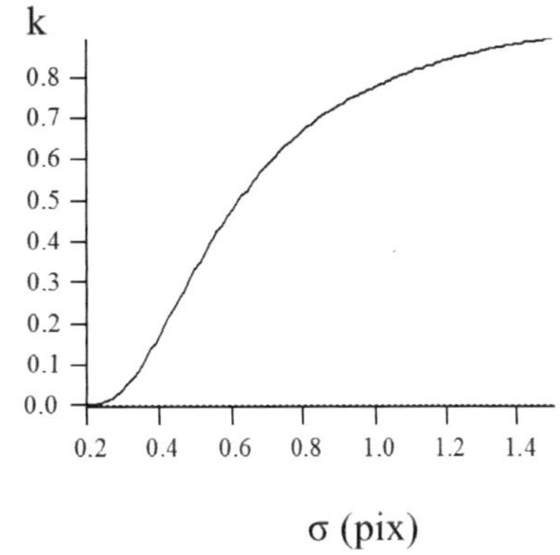

Figure2. Dependency of scale factor k on width of smooth kernel σ (equation (12))

2.2 Experimental validation of noise model parameters estimation algorithm

To verify the proposed algorithm, I compared the noise model parameters estimated from time-lapse sequence with parameters obtained from the single frame of the same sequence (Fig.3).

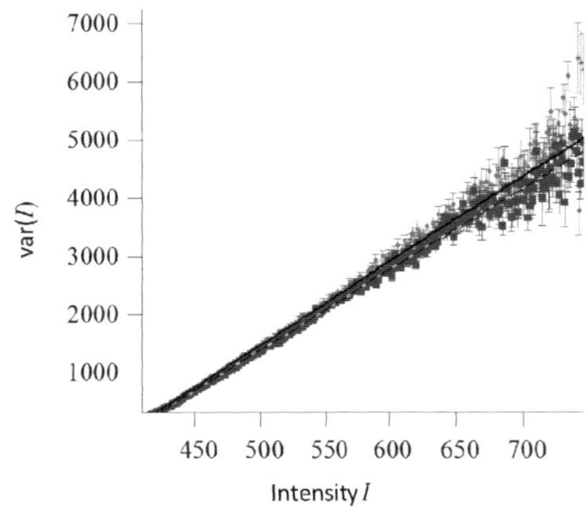

Fig.3. Variance of intensities estimated from the fluctuation of the pixel intensities of time-lapse image sequence (light gray dots) and variance estimation from single frame (dark gray squares). Experimental data presented as mean±SEM. Theoretical model fits are drown by solid and dashed lines. The cells and imaging conditions are the same as in Fig.1. Solid and dashed lines are model fits to time-lapse and single image respectively.

The resulting estimations of variance are overlapped (gray dots and dark gray squares) within error bar (SEM). The parameters of noise modes were derived by maximum likelihood fitting of equation (6) to experimental data and found to be $c = 14.12 \pm 0.02$, $\varepsilon = -5590 \pm 8$ for time-laps and $c = 13.7 \pm 0.03$, $\varepsilon = -5470 \pm 13$ for single image. These two results are fairly close. This method of noise model parameter estimation is incorporated in quantitative microscopy image analysis platform Motiontracking (http://motiontracking.mpi-cbg.de) and routinely used for image analysis [8,9].

3 Conclusions

Correct parametrization of noise model is perquisite for automatic local sensitivity adjustment and threshold tuning for image segmentation [2,4,5] and object classification [8] algorithms as well as fitting theoretical systems biology models to experimental data [9]. In this work I proposed new simple method of Poisson noise model parameter estimation from single snap-shot fluorescence microcopy image. This method does not require microscope calibration, can be performed "off-line" and is applicable for wide variety of Poisson noise dominated microscopy images.

4 References

[1] Waters, J.C., Accuracy and precision in quantitative fluorescence microscopy, JCB, v.185(7), pp.1135-1148, 2009.

[2] Aguet F., Antonescu C.N., Mettlen M., Schmid S.L., Danuser G., Advances in Analysis of Low Signal-to-Noise Images Link Dynamin and AP2 to the Functions of an Endocytic Checkpoint, Develop.Cell, v.26(13), pp.279-291, 2013.

[3] Collinet, C., Stoeter, M., Bradshaw, C.R., Samusik, N., Rink, J.C., Kenski, D., Habermann, B., Buchholz, F., Henschel, R., Mueller, M.S., Nagel, W.E., Fava, E., Kalaidzidis, Y., Zerial, M., Systems Survey of Endocytosis by Multiparametric Image Analysis, Nature, 2010, v.464, pp.243-250, 2010.

[4] van Kempen, G.M.P., van Vliet, L.J., Verveer, P.J., Voort, H.T.M.V.D., A quantitative comparison of image restoration methods for confocal microscopy, J. Microscopy, v.185(3), pp.354-365, 1997.

[5] Rink, J.C., Ghigo, E., Kalaidzidis, Y.L., Zerial, M., Rab Conversion as a Mechanism of Progression from Early to Late Endosomes, Cell, v.122, pp.735-749, 2005.

[6] Lefkimmiatis, S., Papandreou, G., Bayesian Inference on Multiscale Models for Poisson Intensity Estimation: Applications to Photon-Limited Image Denoising, IEEE Transactions On Image Processing, v.18(8), pp.1724-1741, 2009.

[7] Jin, X., Xu, Z., Hirakawa, K., Noise Parameter Estimation for Poisson Corrupted Images Using Variance Stabilization Transforms, IEEE Transactions On Image Processing, v.23(3), pp.1329-1339, 2014.

[8] Morales-Navarrete, H., Segovia-Miranda, F., Klukowski, P., Meyer, K., Nonaka, H., Marsico, G., Chernykh, M., Kalaidzidis, A., Zerial, M., Kalaidzidis, Y., A versatile pipeline for the multi-scale digital reconstruction and quantitative analysis of 3D tissue architecture, eLife, v.4, e11214, 2015.

[9] Meyer, K., Ostrenko, O., Bourantas, G., Morales-Navarrete, H., Porat-Shliom, N., Segovia-Miranda, F., Nonaka, H., Ghaemi, A., Verbavatz, J.M., Brusch, L., Sbalzarini, I., Kalaidzidis, Y., Weigert, R., Zerial, M., A Predictive 3D Multi-Scale Model of Biliary Fluid Dynamics in the Liver Lobule, Cell Systems, v.4(3), pp.277-290, 2017.

SESSION

LATE PAPERS - COMPUTATIONAL BIOLOGY

Chair(s)

TBA

Towards a Method for the Assessment of Cerebral Arteriovenous Malformations Surgery with a Bi-Directional Doppler System for Blood Flow Measurement

E. Rubio-Acosta, D.F. García-Nocetti, P. Acevedo Contla, M. Fuentes-Cruz, J.A. Contreras-Arvizu

Instituto de Investigaciones en Matemáticas Aplicadas y en Sistemas
Universidad Nacional Autónoma de México
Circuito Escolar, Cd. Universitaria, México City, 04510, México
fabian.garcia@iimas.unam.mx

ABSTRACT:
This paper shows the application of Doppler flow measurement techniques to assess the development of surgical treatment of cerebral arteriovenous malformations during the intervention. Also, it is shown the architecture of a Bi-directional Doppler System for Blood Flow Measurement that has already been used to evaluate the bypass quality in coronary revascularization surgeries. Four methods of spectral analysis are described, which include the analysis of the real Doppler signal, its analytic signal and its quadrature signal, using Fourier transform, Gabor transform and time-frequency distributions of the Cohen class, as well as algorithms of flow separation.

KEYWORDS:
Doppler Blood Flow Measurement, Velocity Waveforms Assessment, Cerebral Arteriovenous Malformations, Time Frequency Analysis.

1. Introduction

This paper shows the application of Doppler flow measurement techniques to assess the development of surgical treatment of cerebral arteriovenous malformations during the intervention.

The Bidirectional Doppler System for Blood Flow Measurement shown in this paper has already been successfully used to evaluate the bypass quality in coronary revascularization surgeries. [22][23][24][25][26][27][28][29]. Figure 18 shows a photograph of the prototype of the system used during surgeries. Figure 19 shows a photograph of the system screen assessing a coronary bypass.

It is now intended to use this system to assess the development of surgeries for removal of cerebral arteriovenous malformations.

2.- Arteriovenous Malformations (AVM)

An arteriovenous malformation (AVM) is an abnormal defective formation of blood vessels that drains arterial blood directly into the veins without passing through capillaries [1]. Consequently, the partial lack of blood flow with oxygen in the capillaries can cause tissue damage in the affected areas. This phenomenon of partial lack of blood supply to the tissues caused by an AVM is known as vascular thievery [2] [3] [4].

Figure 1 shows a diagram of a healthy network of capillaries, which distributes oxygenated arterial blood to the surrounding tissues adequately. Figure 2 shows an outline of an arteriovenous malformation that drains oxygenated arterial blood directly into the veins, which causes a significant decrease in the supply of oxygenated arterial blood to the network of capillaries, degrading their function.

Other treatments against AVM include conservative treatments (non-invasive treatments) and surgical treatments (invasive treatments). The latter include surgical removal (resection), endovascular embolization and stereotactic radiosurgery[5].

This paper shows the application of Doppler flowmetry techniques to evaluate the development of surgical treatment during the intervention. The pre-intervention blood flow condition may be as follows: there is a stolen high flow in both the AVM feeding artery and the vein that drains the AVM (points 2 and 3 in figure 2), and an abnormal low flow in the artery that feeds healthy capillaries (point 1 in Figure 2). During the procedure, the stolen blood flow through the artery feeding the AVM is partially or totally blocked (point 2 in figure 2). The post-intervention blood flow condition is low flow (or non-flow) in both the AVM feeding artery and the AVM draining vein (points 2 and 3 in Figure 2), and a restored high flow in the artery feeding the healthy capillaries (point 1 in Figure 2).

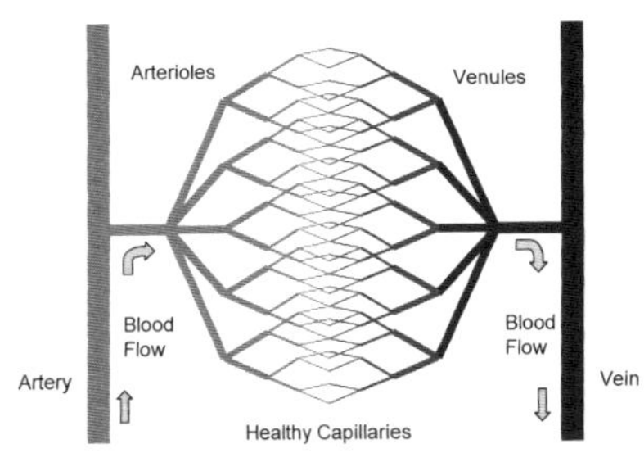

Figure 1: Network of healthy capillaries.

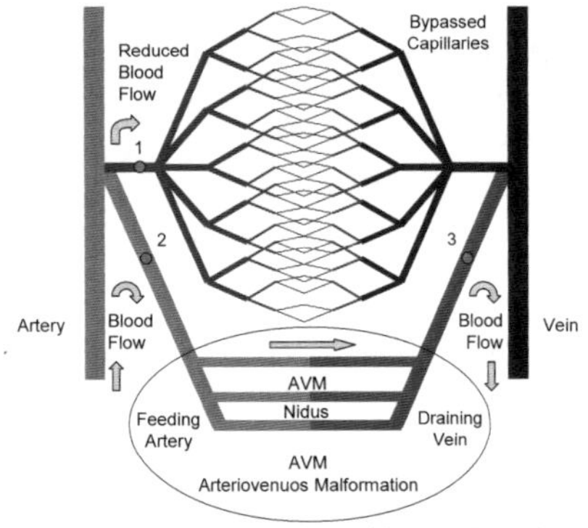

Figure 2: Arteriovenous malformation that robs oxygenated arterial blood to the network of capillaries. Points 1, 2 and 3 in the figure are the points of measurement of blood flow.

3.- Bidirectional Doppler System for Measurement of Blood Flow

The basis of the Doppler Flow measurement applied to blood flow is that the instantaneous mean frequency of the Doppler signal is proportional to the instantaneous mean velocity of the blood flow through the artery or vein [6] [7].

$$f_{Doppler} = \frac{2 f_0 \cos(\alpha)}{c} V_{BloodFlow} \qquad (1)$$

Hence, one of the main objectives of the Bi-Directional Doppler System for Blood Flow Measurement is to estimate the instantaneous mean frequency of the Doppler signal.

The Bi-Directional Doppler System for Blood Flow Measurement consists of three modules. See figure 3. The first module is hardware. This basically consists of the blood flow detector probe, the transducer and the electronic devices that deliver a Quadrature Doppler signal in (two channels: Doppler signal -D signal- and Doppler signal in Quadrature -Q signal-).

The second module is software. This basically consists of a collection of spectral analysis programs whose purpose is to contribute to the estimation of the spectrogram and the instantaneous mean frequency of the Doppler signal.

The third module basically consists of the adequate graphical display of the results produced in the spectral estimation module.

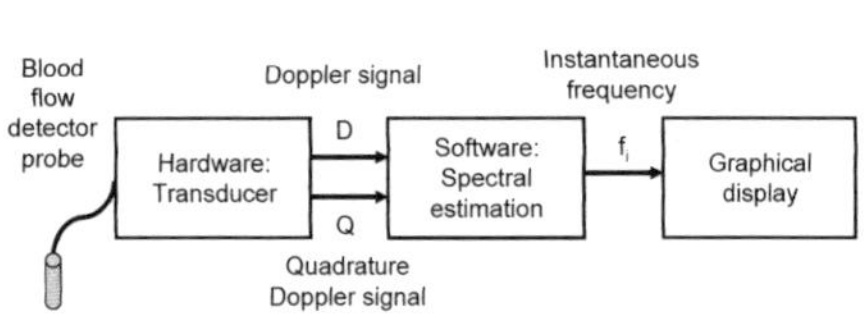

Figure 3: Block diagram of the Bi-Directional Doppler System for Blood Flow Measurement.

4.- Spectral Analysis

Using hardware, the Bi-Directional Doppler System for Blood Flow Measurement generates two input signals to be analyzed by the spectral estimation module: the first is a Doppler signal (D), and the second is another Doppler signal but in Quadrature (Q) with respect to D signal.

These signals are represented by time series of real numbers sampled according to a certain sampling frequency, and their spectral analysis is performed by processing a succession of consecutive data windows, of a certain length, with or without overlap. As a result of processing the succession of data windows, a spectrogram is obtained from which the instantaneous mean frequency is estimated. See figure 4.

There are different options for spectrally analyzing the Doppler D and Q signals. Four of these are explored in this paper.

The four options that will be explained are: the use of Short Time Fourier Transform (STFT) or Gabor Transform to analyze the Doppler signal, the use of Time Frequency Distributions to analyze the analytic signal corresponding to the Doppler signal, the use of Flow Separation Algorithms to analyze the Quadrature Doppler signal, and the use of Time Frequency Distributions to analyze the Quadrature Doppler signal.

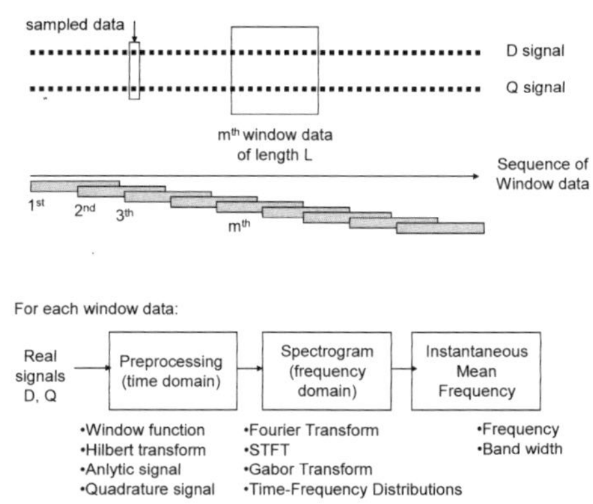

Figure 4. Block diagram of the spectral analysis of a data window.

4.a.- STFT or Gabor Transform

In this method unidirectional flow is assumed (otherwise, if there is bidirectional flow, this option is not adequate). Each data window corresponding to the D signal (real signal) is analyzed consecutively. First, its STFT [8] or its Gabor transform [9] is calculated. Then, its spectrogram is calculated. Finally, as a result, the instantaneous mean frequency is calculated. See Figure 5.

The Short Time Fourier Transform (STFT) of a signal $x(t)$ is defined as:

$$STFT\{x(t)\} \equiv \widehat{X}_{STFT}(t,\omega) = \int_{-\infty}^{\infty} W(\tau - t)x(\tau)e^{-j\omega\tau}\,d\tau \quad (2)$$

where $W(t)$ is a window function; for example: rectangular, Hanning, Kaiser, etc.

The Gabor Transform of a signal $x(t)$ is defined as:

$$G_\alpha\{x(t)\} \equiv \widehat{X}_{G\alpha}(t,\omega) = \int_{-\infty}^{\infty} e^{-\pi\alpha(\tau-t)^2} x(\tau)e^{-j\omega\tau}\,d\tau \quad (3)$$

where α is a parameter that optimizes the tradeoff between time-frequency resolution. Comparing (1) with (2), it is observed that $W(t) = e^{-\pi\alpha t^2}$ is a window function of Gaussian type.

For the transformations (2) and (3), the spectrogram is defined respectively as:

$$spec(t,\omega) = \left|\widehat{X}_{STFT}(t,\omega)\right|^2 \quad (4)$$

$$spec(t,\omega) = \left|\widehat{X}_{G\alpha}(t,\omega)\right|^2 \quad (5)$$

Finally, for this method, the instantaneous mean frequency is defined as:

$$f_i(t) = \frac{\int_0^\infty \omega \bullet spec(t,\omega)\,d\omega}{\int_0^\infty spec(t,\omega)\,d\omega} \quad (6)$$

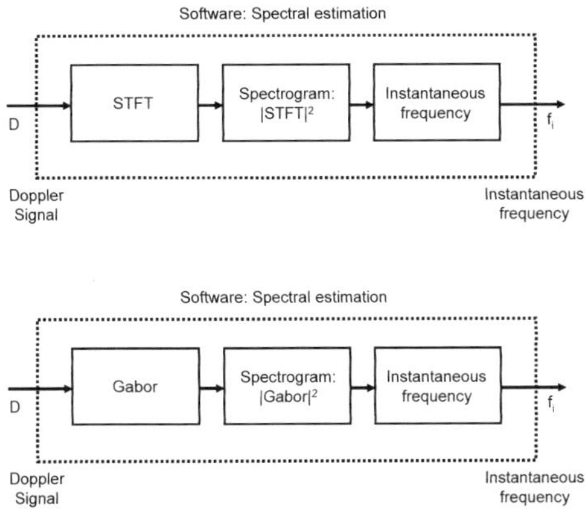

Figure 5: Spectral estimation for unidirectional flow, using the Doppler signal D (real signal) with the Short Time Fourier Transform, or with the Gabor Transform.

4.b.- Analytic Signal with Time-Frequency Distributions

Also, unidirectional flow is assumed in this method. Each data window that corresponds only to D signal (a real signal) is analyzed consecutively, ignoring the Q signal. First, its analytic signal which is a complex signal, is calculated. The analytic signal calculation can be made in time domain, or in frequency domain. In time domain, it is calculated by the Hilbert transform of the D signal (using a convolution); while in frequency domain, its spectrum is constructed using the Fourier transform of the D signal. Then some time-frequency distribution of Cohen's class [10][11] of the analytical signal is calculated (it may also be the spectrogram). Finally, as a result, the instantaneous mean frequency is calculated. See Figure 6.

The analytic signal calculation in time domain is described below. The analytic signal of a real signal $x_r(t)$ is defined as:

$$x_a(t) = x_r(t) + jH\{x_r(t)\} \quad (7)$$

where the H operator means the Hilbert transform. The Hilbert Transform of a signal $x(t)$ is defined as:

$$H\{x(t)\} = \frac{1}{\pi}\int_{-\infty}^{\infty} \frac{x(\tau)}{t-\tau}\,d\tau \quad (8)$$

Note that the Hilbert transform can be calculated by convolution:

$$H\{x(t)\} = h(t) * x(t) \quad (9)$$

where $h(t) = 1/(\pi t)$.

Now, the analytic signal calculation in frequency domain is described below. The Fourier transform of a signal $x(t)$ is defined as:

$$F\{x(t)\} \equiv \widehat{X}(\omega) = \int_{-\infty}^{\infty} x(t)e^{-j\omega t}\,dt \quad (10)$$

and the Inverse Fourier Transform of $\widehat{X}(\omega)$ is defined as:

$$F^{-1}\{\widehat{X}(\omega)\} \equiv x(t) = \frac{1}{2\pi}\int_{-\infty}^{\infty} \widehat{X}(\omega)e^{j\omega t}\,d\omega \quad (11)$$

Then, let $\widehat{X}_r(\omega)$ be the Fourier transform of a real signal $x_r(t)$. The Fourier transform of the analytic signal is constructed as:

$$\widehat{X}_a(\omega)=\widehat{X}_r(\omega)+\operatorname{sign}(\omega)\widehat{X}_r(\omega)=$$

$$=\begin{cases}2\widehat{X}_r(\omega) & ,\omega>0\\ \widehat{X}_r(\omega) & ,\omega=0\\ 0 & ,\omega<0\end{cases} \quad (12)$$

where $\operatorname{sign}(\omega)$ is the sign function. The analytic signal $x_a(t)$ is the inverse Fourier transform of $\widehat{X}_a(\omega)$.

On the other hand, the Time-Frequency Distributions (TFD) of the Cohen Class are defined as:

$$TFD_\phi(t,\omega)=$$

$$=\frac{1}{4\pi^2}\int_{-\infty}^{+\infty}\int_{-\infty}^{+\infty}\int_{-\infty}^{+\infty}x\left(\mu+\frac{\tau}{2}\right)x^*\left(\mu-\frac{\tau}{2}\right)\bullet \quad (13)$$

$$\bullet\ \phi(\theta,\tau)e^{-j\theta t+j\theta\mu-j\omega\tau}d\theta d\mu d\tau$$

where $\phi(\theta,\tau)$ is the distribution kernel, which determines its properties. Table 1 shows the kernel of some of the major TFD.

Finally, for this method, the instantaneous mean frequency [14][15] is defined as:

$$f_i(t)=\frac{\int_0^\infty \omega\bullet TFD_\theta(t,\omega)d\omega}{\int_0^\infty TFD_\theta(t,\omega)d\omega} \quad (14)$$

TFD	Kernel $\phi(\theta,\tau)$
Wigner-Ville [16][17][18][19]	1
Choi-Williams [20]	$e^{-\frac{\theta^2\tau^2}{\sigma}}$
Modified-B [21]	$\frac{\Gamma(\alpha+j\pi\theta)\Gamma(\alpha-j\pi\theta)}{\Gamma(\alpha^2)}$
Spectrogram Eq. (3)	$\int_{-\infty}^{\infty}W(u+\frac{1}{2}\tau)W^*(u-\frac{1}{2}\tau)e^{-j\theta u}du$

Table 1. Kernel of some time frequency distributions of the Cohen class (α and σ are parameters that optimize the time-frequency resolution).

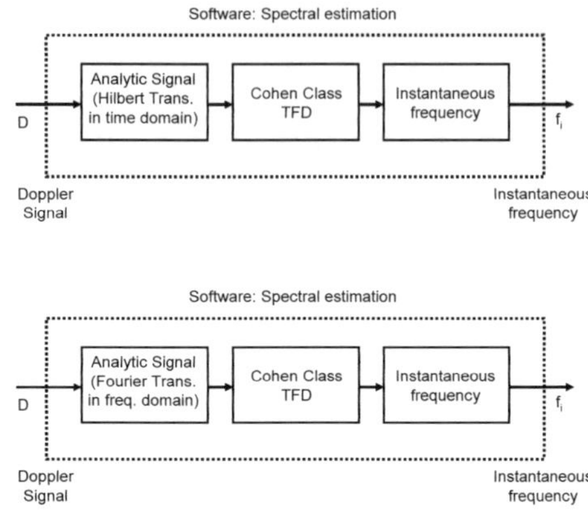

Figure 6. Spectral estimation for unidirectional flow, using the analytic signal of the Doppler signal D (a complex signal) with the time-frequency distributions of the Cohen class. The construction of the analytic signal is performed with the Hilbert transform in time domain or with the Fourier transform in frequency domain.

4.c.- Quadrature Signal with Flow Separation Algorithm

This method assumes bidirectional flow (note that unidirectional flow is a particular case). Each data window corresponding to the quadrature signal $D(t)+jQ(t)$, which is a complex signal, is analyzed consecutively. First, a flow separation algorithm [12] [13] is applied to generate two real signals: one forward flow signal $x_{forward}(t)$ and one reverse flow signal $x_{reverse}(t)$. The flow separation algorithm involves calculating the Hilbert transform of the D signal. Finally, each unidirectional flow signal $x_{forward}(t)$ and $x_{reverse}(t)$ is analyzed separately according to the first or second options set forth above. See Figure 7.

The flow separation algorithm is:

$$x_{forward}(t)=Q(t)-H\{D(t)\} \quad (15)$$

$$x_{reverse}(t)=Q(t)+H\{D(t)\} \quad (16)$$

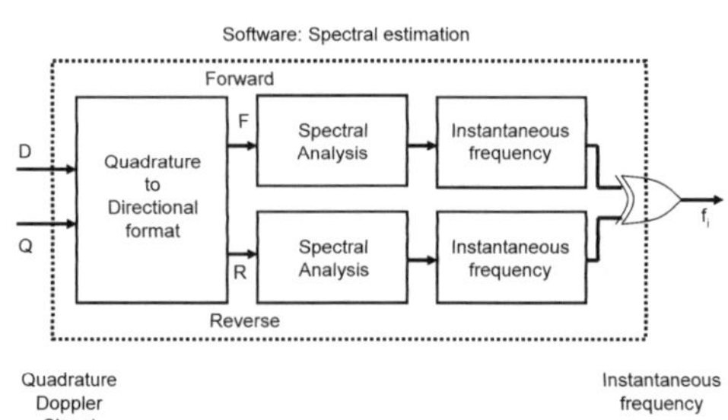

Figure 7. Spectral estimation for bidirectional flow using the quadrature Doppler signal (complex signal) and a flow separation algorithm. The result of this algorithm are two

real signals, one of forward flow (F) and one of reverse flow (R), whose spectral analyzes are done separately.

4.d.- Quadrature Signal with Time-Frequency Distributions

This method also assumes bidirectional flow. Each data window corresponding to the quadrature signal $D(t)+jQ(t)$, which is a complex signal, is analyzed consecutively. First, some time-frequency distribution of the Cohen class of the quadrature signal (13) is calculated (it can also be the spectrogram). Finally, as a result, the instantaneous mean frequency is calculated. See figure 8.

For this method, the instantaneous mean frequency is defined as:

$$f_i(t)=\frac{\int_{-\infty}^{\infty}\omega\bullet TFD_\theta(t,\omega)d\omega}{\int_{-\infty}^{\infty}TFD_\theta(t,\omega)d\omega} \quad (17)$$

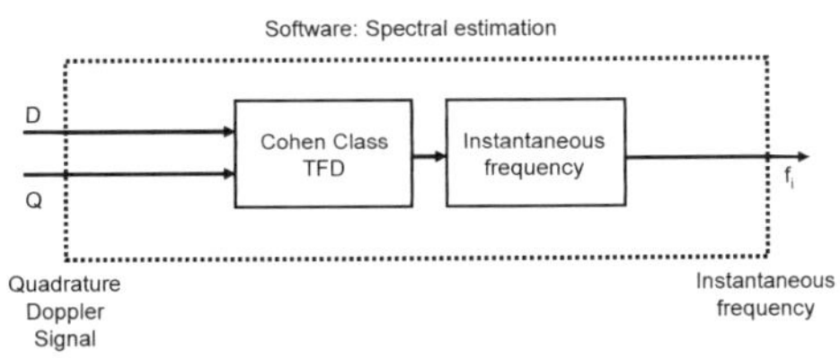

Figure 8. Spectral estimation for bidirectional flow using the quadrature Doppler signal (complex signal) and the time-frequency distributions of the Cohen class.

5.- Results

The results are presented in two sections. The first section consists of a qualitative comparison of the methods. The second section consists of a preliminary assessment of a surgical procedure to remove a cerebral arteriovenous malformation.

The amplitude of the spectrograms is plotted on a logarithmic scale, normalized with the maximum expected power in each case. A range of 0 [dB] to -8 [dB] is plotted.

Note that the nature of the flow signal being analyzed is unidirectional because they correspond to flow signals in cerebral arteries. Conventionally it is said to be a forward flow signals, which have no reverse flow components. When a complex flow signal (such as the analytic signal or the quadrature signal) is spectrally analyzed, the positive frequency components are conventionally associated with the forward flow and the negative frequency components with the reverse flow.

5.a.- Qualitative comparison of methods

This section only shows results of the spectral analysis of the artery that feeds the capillaries with its restored flow. See figure 15.

Figure 9 shows the spectrogram of the signal D (real signal) processed with STFT. Figure 10 shows the spectrogram of the D signal (real signal) processed with Gabor transform (method 4.a). Since a real signal is analyzed in both cases, the unidirectional flow requirement is requested. If the flow were bidirectional, the information corresponding to the forward and reverse flow would be inseparably mixed in the real signal. Also, since a real signal is analyzed, the spectrogram is symmetrical. So the same information contained in the positive frequencies is in the negative frequencies of the spectrogram. For this reason, the integral to calculate the instantaneous mean frequency (6) is over the interval [0,ℕ). Negative frequencies are thus ignored.

The spectrogram of the analytic signal of the D signal processed with time frequency distributions (method 4.b) is shown in Figure 11. The analytic signal is a complex signal. Since the spectrogram of a complex signal is calculated, the spectrogram is not symmetrical. So the information contained in the positive and negative frequencies of the spectrogram is different. However, since the original signal being analyzed is the D signal, which is a real signal, the unidirectional flow requirement is requested. Consequently, only the positive frequencies have information and the negative frequencies do not, canceling out. For this reason, the integral to calculate the instantaneous mean frequency (14) is over the interval [0,ℕ).

Figure 12 shows the spectrograms of forward and reverse flow signals processed with STFT. The flow separation algorithm is applied to the quadrature Doppler signal (D+jQ), which is a complex signal. So the information corresponding to the forward and reverse flow is mixed but separable in the complex signal. However, since the nature of the flow signal being analyzed is unidirectional, only the spectrogram associated with the forward flow has information, whereas the spectrogram associated with the reverse flow does not, canceling out.

Figure 13 shows the spectrogram of the Quadrature Doppler signal processed with time frequency distributions (method 4.d). The Quadrature Doppler signal (D+jQ) is a complex signal. Immediately, the positive frequencies of the spectrogram are related to forward flow and the negative frequencies of the spectrogram are related to reverse flow. For this reason, the integral to calculate the instantaneous mean frequency (17) is over the interval (-ℕ,ℕ). The fact that the negative frequencies of the spectrogram are zero was explained in the previous paragraph.

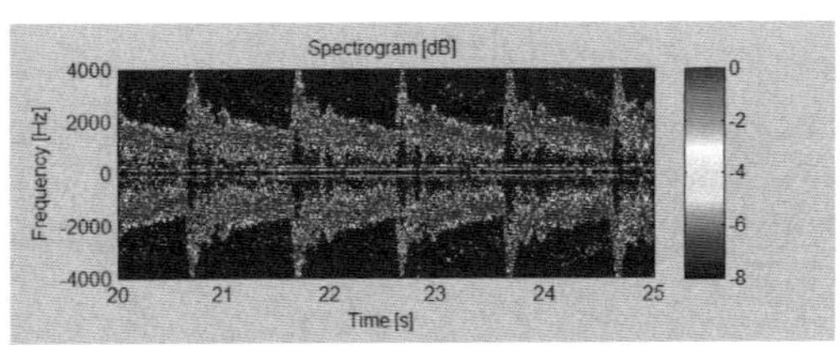

Figure 9.- Spectrogram calculated with Doppler D signal (real signal) and STFT (method 4.a).

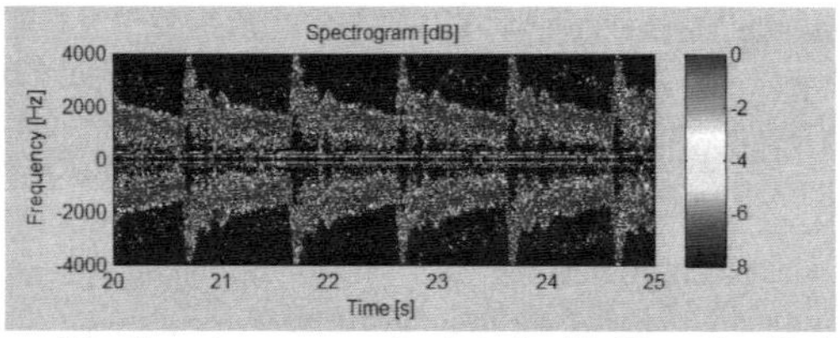

Figure 10.- Spectrogram calculated with Doppler D signal and Gabor transform (method 4.a).

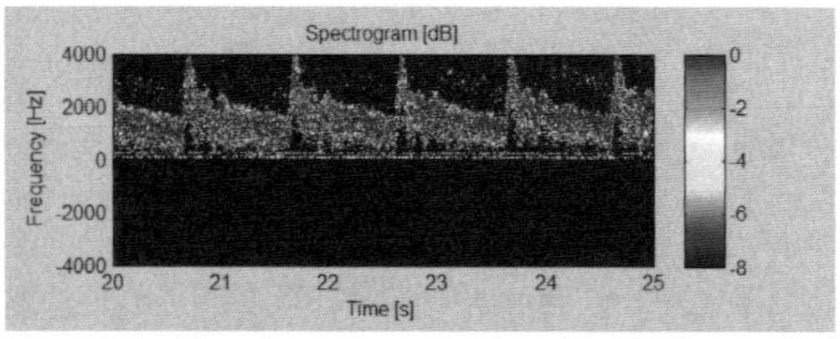

Figure 11.- Spectrogram calculated with the analytic signal of the Doppler D signal and time frequency distributions (method 4.b).

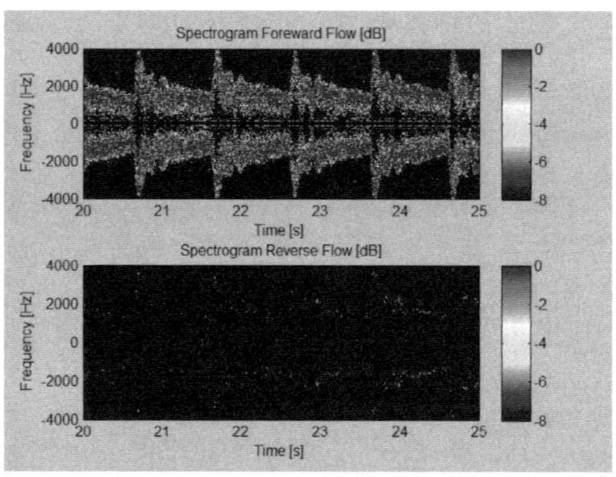

Figure 12. Spectrograms calculated with the Doppler signal in quadrature (D+jQ) applying a flow separation algorithm (forward flow and reverse flow) and time frequency analysis to each of the mentioned flows (method 4.c).

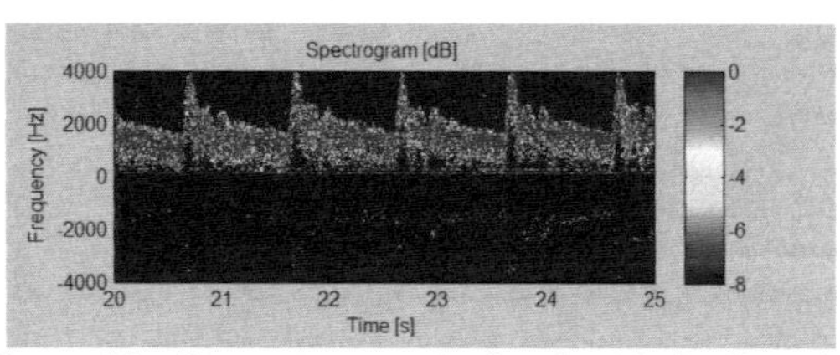

Figure 13.- Spectrogram calculated with the Doppler signal in quadrature (D+jQ) and time frequency distributions (method 4.d).

5.b.- Preliminary assessment of a surgical procedure to remove cerebral arteriovenous malformation

Figures 14 to 17 show three boxes. The upper box corresponds to the spectrogram of the signal; the middle box corresponds to the instantaneous mean frequency of the signal (proportional to the instantaneous mean velocity of the flow); and the box below corresponds to the electrocardiogram, which is used as a reference.

As explained in section 2, the flow conditions before surgical intervention are: the presence of a low flow in the feeding artery to the capillaries due to vascular theft by the arteriovenous malformation, and the presence of a high flow in the draining vein from the arteriovenous formation. The first situation is shown in figure 14 (measurement made in point 1 of figure 2) and the second in figure 16 (measurement made in point 3 of figure 2).

On the other hand, the flow conditions after the surgical intervention are: the presence of a restored high flow in the feeding artery of the capillaries due to the gradual extirpation of the arteriovenous malformation, and the presence of a low flow in the draining vein from the arteriovenous formation. The first situation is shown in figure 15 (measurement made in point 1 of figure 2) and the second in figure 17 (measurement made in point 3 of figure 2).

Table 2 shows the average values of the instantaneous mean frequencies (proportional to the instantaneous mean velocity of the flow), of the feeding artery to capillaries and of the draining vein of the arteriovenous malformation, before and during surgery.

	Before surgery	During surgery
Artery of Capillaries	829 [Hz] Low Flow	1683 [Hz] High Flow
Vein of AVM	1343 [Hz] High Flow	1221 [Hz] Low Flow

Table 2. Average values of the instantaneous mean frequencies (proportional to the instantaneous mean velocity of the flow), of the feeding artery to capillaries and of the draining vein of the arteriovenous malformation, before and during surgery.

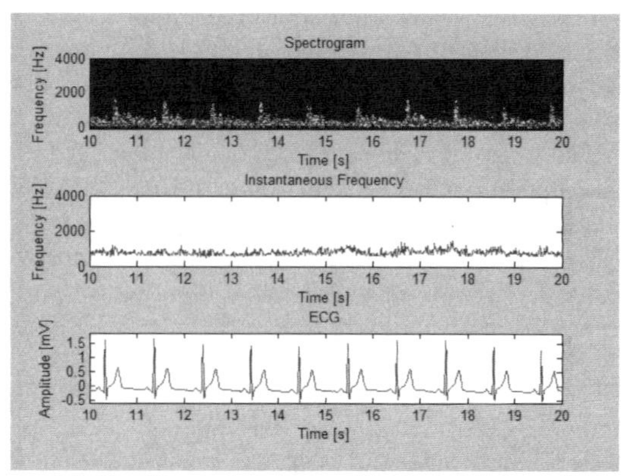

Figure 14. Feeding artery to capillaries with low flow due to vascular theft by arteriovenous malformation. Condition shown before initiating the removal of arteriovenous malformation.

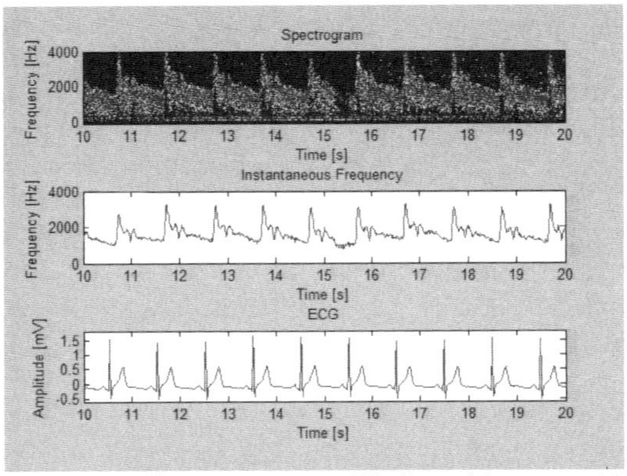

Figure 15. Feeding artery to capillaries with restored high flow due to gradual removal of arteriovenous malformation.

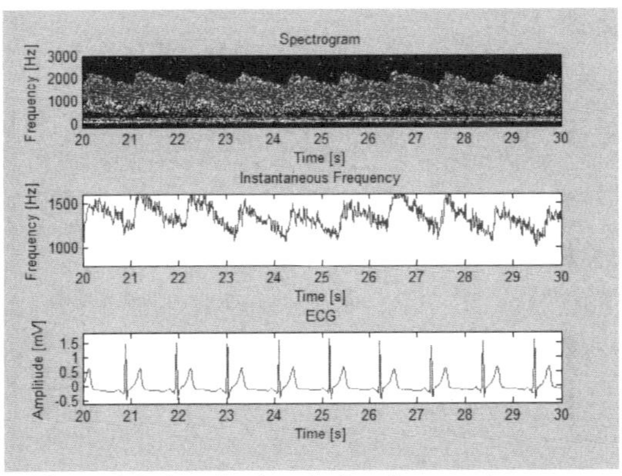

Figure 16. A vein that drains the arteriovenous malformation with high flow. Condition shown before initiating the removal of arteriovenous malformation.

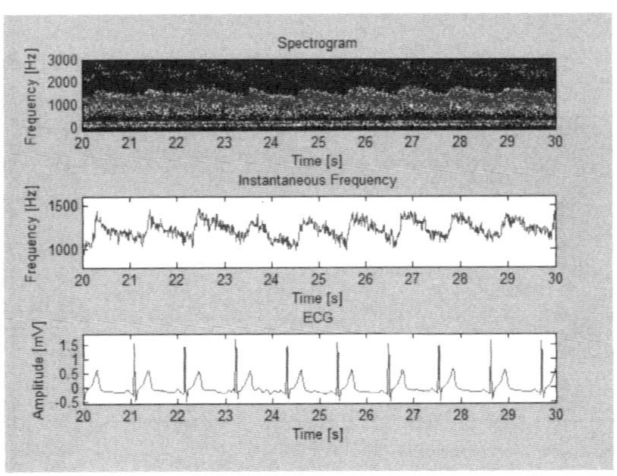

Figure 17. A vein that drains arteriovenous malformation with low flow. Condition shown during the removal of arteriovenous malformation.

6.- Conclusions

This paper shows the application of Doppler flow measurement techniques to assess the development of surgical treatment of cerebral arteriovenous malformations during the intervention.

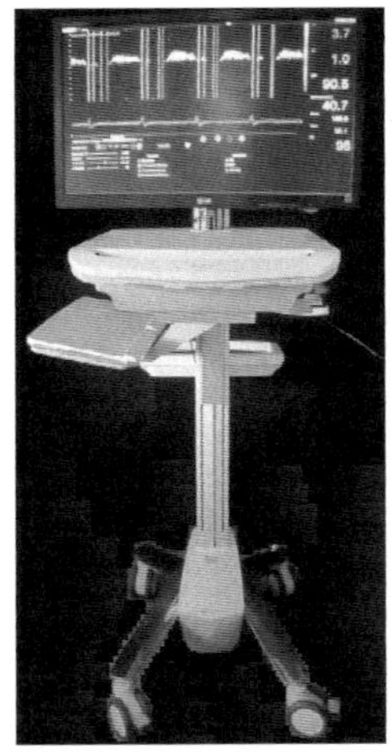

Figure 18. Prototype of the Bi-Directional Doppler System for Measurement of Blood Flow used during surgeries.

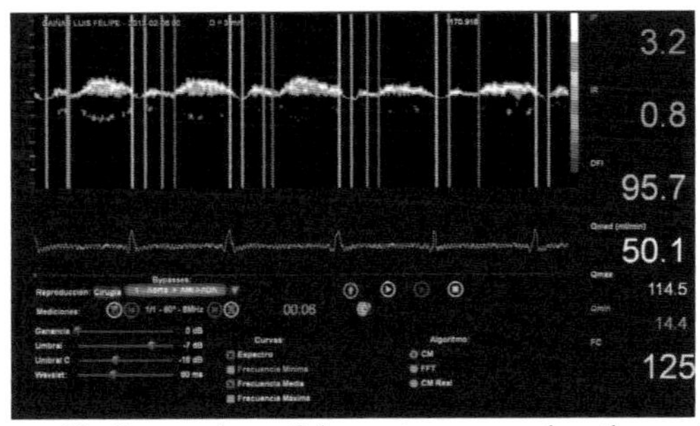

Figure 19. Screenshot of the system assessing the quality of a coronary bypass.

Also, it is shown the architecture of a Bi-directional Doppler System for Blood Flow Measurement that has already been used to evaluate the bypass quality in coronary revascularization surgeries.

Four suitable spectral analysis methods are described: the use of Short Time Fourier Transform (STFT) or Gabor Transform to analyze the Doppler signal, the use of Time Frequency Distributions to analyze the analytic signal corresponding to the Doppler signal, the use of Flow Separation Algorithms to analyze the Quadrature Doppler signal, and the use of Time Frequency Distributions to analyze the Quadrature Doppler signal.

Acknowledgements

The authors acknowledge projects DGAPA-UNAM (PAPIIT-IT101316) and (PAPIIT-IT106016) for their financial support. Also we want to acknowledge E. Nathal and K. Pedroza for providing signals for the development of this work.

References

[1] Term "Arteriovenous Malformations, (D001165)" in NIH National Library of Medicine. https://meshb.nlm.nih.gov/record/ui?ui=D001165

[2] Lopez G., et al, "Malformaciones arteriovenosas cerebrales: Desde el diagnóstico, sus clasificaciones y patofisiología, hasta la genética", Revista Mexicana de Neurociencia, Vol. 11, No. 6, 2010, pp 470-479

[3] Fernández-Melo R., et al, "Diagnóstico de las malformaciones arteriovenosas cerebrales". Revista de Neurología, Vol. 37, No. 9, 2003, pp 870-878

[4] López-Flores G., et. al. "Etiopatogenia y fisiopatología de las malformaciones arteriovenosas cerebrales" Archivos de Neurociencias (Mex), Vol. 15, No. 4, 2010, pp 252-259

[5] Martínez-Ponce de León A., et al, "Malformaciones arteriovenosas cerebrales: evolución natural e indicaciones de tratamiento", Medicina Universitaria Vol. 11, No. 42, 2011, pp 44-54

[6] Fish P., "Physics and Instrumentation of Diagnostic Medical Ultrasound", Ed. Wiley, 1991.

[7] Evans D., McDicken N., "Doppler Ultrasound: Physics, Instrumentation and Signal Processing", Ed. Wiley, 2nd edition, 2000.

[8] Oppenheim A., Schafer R., "Discrete-Time Signal Processing", Prentice Hall, 3rd edition, 2009.

[9] Gabor, "Theory of communication. Part 1: The analysis of information", Journal of the Institution of Electrical Engineers - Part III: Radio and Communication Engineering, Vol. 93, No. 26, 1946, pp 429 - 441

[10] Cohen L. "Time-Frequency Distributios: A Review". Proceedings of the IEEE, Vol. 77, No. 7, 1989, pp 941-981

[11] Cohen L. "Time-Frequency Analysis. Theory and Applications", Prentice Hall PTR, 1994.

[12] Aydin N., Fan L., Evans D., "Quadrature-to-directional format conversion of Doppler signals using digital methods", Physiological measurement Vol. 15, 1994, pp 181-199

[13] N Aydin, DH Evans, "Implementation of directional Doppler techniques using a digital signal processor", Medical and Biological Engineering and Computing, Vol. 32, pp 157-164

[14] Boashash B., "Estimating and Interpreting The Instantaneous Frequency of a Signal -Part 1: Fundamentals", Proceedings of the IEEE, Vol. 80, No. 4, 1992, pp 520-538

[15] Boashash B., "Estimating and Interpreting the Instantaneous Frequency of a Signal -Part 2: Algorithms and Applications", Proceedings of the IEEE, Vol. 80, No. 4, 1992, pp 540-568

[16] Claasen T.A.C.M and Mecklenbräuker W.F.G., "The Wigner distribution - A tool for time-frequency signal analysis. Part I: Continuous-time signals", Philips J. Res. Vol. 35, 1980, pp 217-250

[17] Claasen T.A.C.M and Mecklenbräuker W.F.G., "The Wigner distribution - A tool for time-frequency signal analysis. Part II: Discrete-time signals", Philips J. Res. Vol. 35, 1980, pp 276-300

[18] Claasen T.A.C.M and Mecklenbräuker W.F.G., "The Wigner distribution - A tool for time-frequency signal analysis. Part III: Relations with other time-frequency signal transformations", Philips J. Res. Vol. 35, 1980, pp 372-389

[19] Martin W., Flandrin P., "Wigner-Ville Spectral Analysis of Nonstationary Processes", IEEE Transactions on Acoustics, Speech, and Signal Processing, Vol. ASSP-33, No. 6, 1985, pp 1461-1470

[20] Choi H. I., Williams W. J., ''Improved time-frequency representation of multicomponent signals using exponential kernels,'' IEEE Transactions on Acoustics, Speech, and Signal Processing, Vol. 37, No. 6, 1989, pp 862–871

[21] B. Barkat B., Boashah B., "A High-resolution quadratic time-frequency distribution for multicomponent signals analysis", IEEE Transactions on Signal Processing, Vol. 49, No. 10, 2001, pp

[22] García NF, Solano GJ, Rubio AE, Moreno HE. "Parallel Computing in Time-Frequency Distributions for Doppler Ultrasound Blood Flow Instrumentation", Revista Mexicana de Ingeniería Biomédica 2001, Vol. XXII, No. 1, 2001, pp 12-19.

[23] García-Nocetti F., et.al. "Sistema Doppler bidireccional para medición de flujo sanguíneo basado en una arquitectura abierta". Revista Mexicana de Ingeniería Biomédica, Vol. XXIV, No. 2, 2003, pp 135-143

[24] J Solano, M Fuentes, A Villar, J Prohias, F Garcia-Nocetti, "Doppler Ultrasound Blood Flow Measurement System for Assessing Coronary Revascularization", Int'l. Conf. Bioinformatics and Computational Biology, BIOCOMP'11, Las Vegas, USA, 2011.

[25] F. García-Nocetfi, J. Solano González, E. Rubio Acosta, "Improving Performance of a TFD-based Spectral Estimation Method in Doppler Ultrasound Blood Flow Measurement", Int'l. Conf. Bioinformatics and Computational Biology, BIOCOMP'12, Las Vegas, USA, 2012.

[26] F. García-Nocetfi, J. Solano González, E. Rubio Acosta, "Advances in Performance Improvement of Time-Frequency Distributions for Doppler Ultrasound Blood Flow Instrumentation", Int'l. Conf. Bioinformatics and Computational Biology, BIOCOMP'13, Las Vegas, USA, 2013.

[27] F. García-Nocetfi, J. Solano González, E. Rubio Acosta, "Optimal Scaling Values for Time-Frequency Distributions in Doppler Ultrasound Blood Flow' Measurement", Int'l. Conf. Bioinformatics and Computational Biology, BIOCOMP'14, Las Vegas, USA, 2014.

[28] F. García-Nocetfi, J. Solano González, E. Rubio Acosta, "A Proposed Warped Modified-B Time-Frequency Distribution Applied to Doppler Blood Flow Measurement", Int'l. Conf. Bioinformatics and Computational Biology, BIOCOMP'15, Las Vegas, USA, 2015.

[29] F. García-Nocetfi, J. Solano González, E. Rubio Acosta, "A Proposed Warped Choi Williams Time-Frequency Distribution Applied to Doppler Blood Flow Measurement", Int'l. Conf. Bioinformatics and Computational Biology, BIOCOMP'16, Las Vegas, USA, 2016.

HABase: A Web-Application for the Analysis of Protein Spectra and Identification of Microbial Species

Michael LaMontagne[1], Thrishala Shetty[2], Tilak Gajjar[2], Chandana Kayyuru[2], Sachin Sriram[2], Chunlong Zhang[1], and Pradeep Buddharaju[2]

[1]Biological & Environmental Sciences, University of Houston Clear Lake, Houston, TX, USA

[2]Computer Science & Computer Information Systems, University of Houston Clear Lake, Houston, TX, USA

Abstract—*Matrix assisted laser desorption ionization - time of flight analysis (MALDI-TOF) provides a rapid and accurate method of identifying microbes. Microbial identification requires a library of protein spectra to query. Many isolates, particularly from environmental samples and emerging pathogens, do not have corresponding spectra in available databases. Investigators can use cluster analysis to identify microbial species from protein spectra without a database; however, this analysis requires either proprietary software or the use of scripting language. Here we present HABase, a graphical user interface for running MALDIquant an R package that allows cluster analysis of protein spectra generated by MALDI-TOF. This user-friendly application facilitates adjustment of key parameters in protein spectra analysis, including selection of signal to noise ratio and setting the baseline. HABase accepts input of raw spectra and produces two primary outputs: phylogenetic trees that allow users to identify clusters of closely related isolates and a matrix of aligned spectra.*

Keywords: MALDI-TOF, proteomics, microbial identification, web application

1. Introduction

MALDI-TOF MS is a recent life science instrumental technique for high-mass molecules that has been applied to diverse areas, ranging from environmental microbiology [1] to histochemistry [2]. For microbial identification, MALDI-TOF systems compare favorably to conventional methods [3], automated phenotyping systems [4] used in clinical settings and DNA sequencing [5].

Several MALDI-TOF systems are commercially available for microbial identification by pattern matching against reference spectra. This provides strain-level identification of microbes [6], costs pennies per isolate and has the throughput to enable an in-depth characterization of microbial isolates comparable to that of next generation sequencing technology [7]; however, these systems require a rich database of reference spectra. Commercial systems largely contain spectra generated from clinical isolates. This hinders the application of MALDI-TOF systems to identification of microbes isolated from the environment. Users can create custom libraries, but this requires expensive proprietary software.

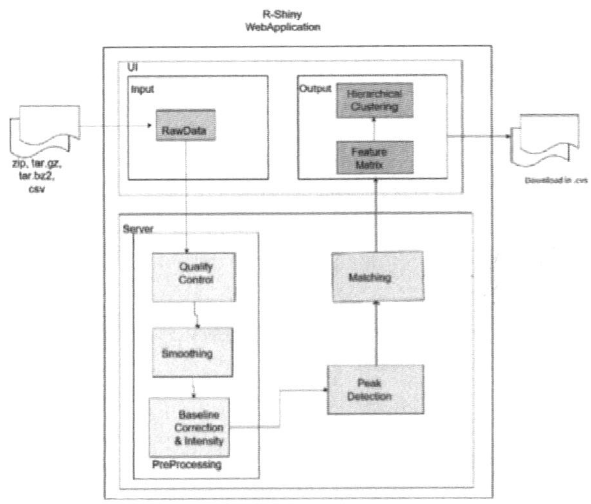

Fig. 1: Architecture Diagram

This cost, and the time associated with building a custom, in-house library, limits the adoption of this technology. Several freeware applications allow analysis of protein spectra. MALDIquant has powerful features, including alignment, peak detection and cluster analysis [8] and runs in R, an open source programming language and software environment for statistical computing and graphics.

As the first step towards developing a public reference database of protein spectra we developed HABase. This web application uses Rshiny to run MALDIquant and allows users to process raw protein spectra generated by MALDI-TOF MS of microbial isolates. The system allows users to adjust data analysis parameters, including signal to noise ratio (SNR), and select various data smoothing algorithms. Users can then export aligned spectra in a matrix format suitable for creating a database of protein spectra.

2. Software Architecture

HABase code has a user interface (UI) and server. UI sets the display screen, including banner titles and tabs for various functions. These functions include Smoothing, Clustering and Peaks. The server code starts when the user uploads spectra. HABase allows users to upload data in these

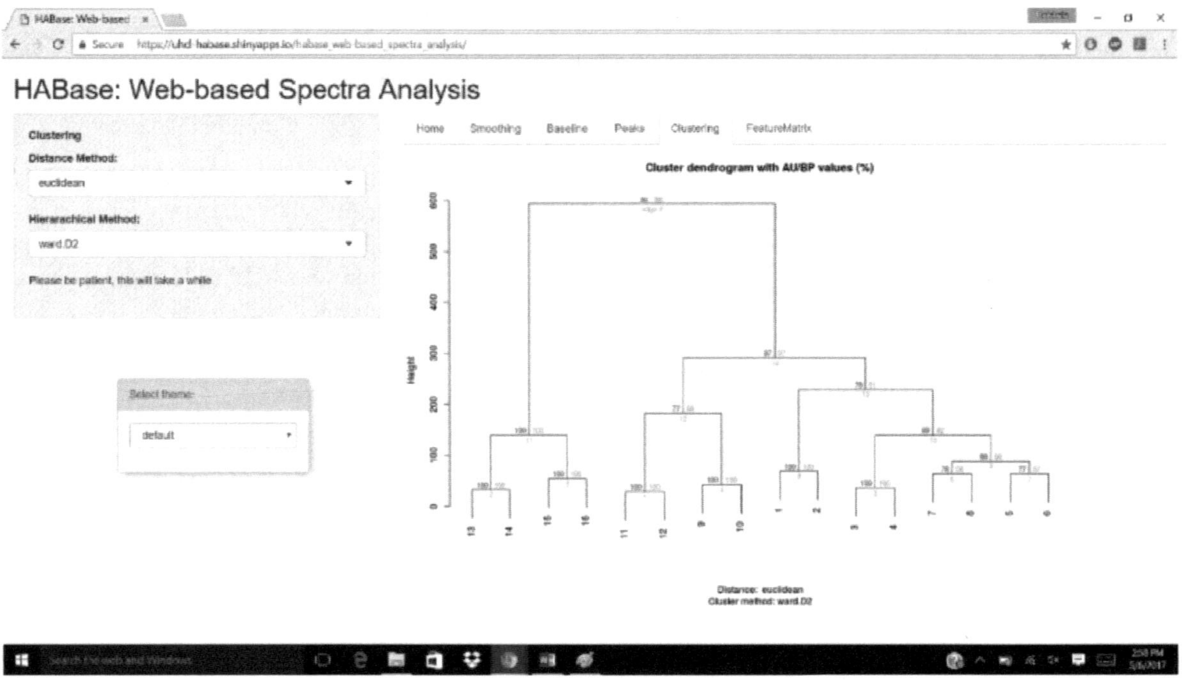

Fig. 2: Screen shot of cluster plot generation in HABase

compressed formats: zip, tar.gz, tar.bz2, and csv. System functions include preprocessing, peak detection and matching (Fig. 1). Preprocessing involves checking for empty or irregular spectra files, displaying the spectrum length and generating plots of individual spectra for visual inspection. Later variance-stabilizing transformation is applied followed by smoothing filter. Further baseline correction is performed on the spectra. At each step plots are developed and viewed in the web application. The application then generates a matrix of aligned spectra. This feature matrix can then be used for cluster analysis, using the R package Pvclust [9], to visualize phylogenetic relationships, and exported in csv format.

This link connects to HABase: `https://uhcl-habase.shinyapps.io/habase_web-based_spectra_analysis/`. The application includes an example data set from Fiedler et al [10]. With this example, data set users can generate a tree showing the relationship between serum samples, as assessed by MALDI-TOF (Fig. 2).

References

[1] Emami K, Nelson A, Hack E, Zhang J, Green DH, and Caldwell GS, "Maldi-tof mass spectrometry discriminates known species and marine environmental isolates of pseudoalteromonas," *Front Microbiol*, vol. 7, no. 104, 2016.

[2] Aichler M and Walch A, "Maldi imaging mass spectrometry: current frontiers and perspectives in pathology research and practice," *Lab Invest*, vol. 95, no. 4, pp. 422–431, 2015.

[3] Barberis C, Almuzara M, Join-Lambert O, Ramírez MS, Famiglietti A, and Vay C, "Comparison of the bruker maldi-tof mass spectrometry system and conventional phenotypic methods for identification of gram-positive rods," *PLoS ONE*, vol. 9, no. 9, 2014.

[4] Saffert RT, Cunningham SA, Ihde SM, Monson Jobe KE, Mandrekar J, and Patel R, "Comparison of bruker biotyper matrix-assisted laser desorption ionization?time of flight mass spectrometer to bd phoenix automated microbiology system for identification of gram-negative bacilli," *Clinical Microbiol*, vol. 49, no. 3, pp. 887–892, 2011.

[5] Barbano D, Diaz R, Zhang L, Sandrin T, Gerken H, and Dempster T, "Rapid characterization of microalgae and microalgae mixtures using matrix-assisted laser desorption ionization time-of-flight mass spectrometry (maldi-tof ms)," *PLoS ONE*, vol. 10, no. 8, 2015.

[6] Singhal N, Kumar M, Kanaujia PK, and Virdi JS, "Maldi-tof mass spectrometry: An emerging technology for microbial identification and diagnosis," *Front Microbiol*, vol. 6, no. 791, 2015.

[7] Lagier J-C, Hugon P, Khelaifia S, Fournier P-E, La Scola B, and Raoult D, "The rebirth of culture in microbiology through the example of culturomics to study human gut microbiota," *Clin Microbiol Rev*, vol. 28, no. 1, pp. 237–264, 2015.

[8] Gibb S and Strimmer K, "Maldiquant: a versatile r package for the analysis of mass spectrometry data," *Bioinformatics*, vol. 28, no. 17, pp. 2270–2271, 2012.

[9] Suzuki R and Shimodaira H, "Pvclust: an r package for assessing the uncertainty in hierarchical clustering," *Bioinformatics*, vol. 22, no. 12, pp. 1540–1542, 2006.

[10] Fiedler GM, Leichtle AB, Kase J, Baumann S, Ceglarek U, and Felix K, "Serum peptidome profiling revealed platelet factor 4 as a potential discriminating peptide associated with pancreatic cancer," *Clin Cancer Res*, vol. 15, no. 11, pp. 3812–3819, 2009.

Author Index

CSREA Press
ATTN: Hamid R. Arabnia
1060 Long Creek Drive
Watkinsville, GA 30677
USA

ISBN 978-1-5108-5386-7

9 781510 853867

CURRAN ASSOCIATES INC.
proceedings.com